William Gee

Short Studies in Nature Knowledge

An Introduction to the Science of Physiography

William Gee

Short Studies in Nature Knowledge
An Introduction to the Science of Physiography

ISBN/EAN: 9783337035471

Printed in Europe, USA, Canada, Australia, Japan

Cover: Foto ©berggeist007 / pixelio.de

More available books at **www.hansebooks.com**

SHORT STUDIES

IN

NATURE KNOWLEDGE

SHORT STUDIES

IN

NATURE KNOWLEDGE

AN INTRODUCTION TO THE SCIENCE
OF PHYSIOGRAPHY

BY

WILLIAM GEE

CERTIFICATED TEACHER OF THE EDUCATION DEPARTMENT, AND
OF THE SCIENCE AND ART DEPARTMENT; ASSISTANT-LECTURER, MANCHESTER
FIELD NATURALISTS' SOCIETY

London

MACMILLAN AND CO.

AND NEW YORK

1895

PREFACE

In these chapters upon Nature Knowledge the author, a teacher of the Education Department, and of the Science and Art Department, has expanded notes of class-lessons given during the past quarter of a century, and has included some of the results of recent travel and research.

The book, whilst introducing the general public to some acquaintance with phases of the natural world, should also serve as a reader and companion to the text-books used in the upper forms of schools, assist University Extension students, aid Kindergarten teachers in preparing for the Froebel Union certificate, and, generally, give an idea of the extended treatment adopted by the great geographers, as Reclus, Mrs. Somerville, Ansted, Malte-Brun, and Carl Ritter.

By adding copious extracts from travellers and eye-witnesses, as Humboldt, Kingsley, and Livingstone, of the scenes they depict, and by dealing with natural phenomena in relation to human affairs, it is sought to increase the attractiveness of the subject.

Footnotes support the text, affording more detailed information for the use of students preparing the study of Physiography for the various public examinations.

The illustrations will also, it is hoped, stimulate to a keener appreciation of natural beauty in the world around us.

The author is much indebted to Mr. C. L. Barnes, M.A. (Oxon.), for valuable suggestions, and for reading the proofs.

MANCHESTER, 1895.

CONTENTS

I

THE GREAT GLOBE ITSELF

Early geographical notions—The spread of earth knowledge, by commerce, by military enterprise—Great discoverers by sea and land : Columbus ; Vasco de Gama ; Magellan — English sea-captains—Land explorers—The earth's physical habitudes —Astronomical relations, surface and extent — Geological character : worlds before the present one — The moulding physical forces : heat, electric power—The future   Page 1

II

MOUNTAINS, VALLEYS, AND GREAT PLAINS

First notions — Highlands of the earth — Breadth of mountain ranges—The Himalayas—The Great Divide of Europe—Mont Blanc—Frosty Caucasus— Tabor—Sinai—American mountains —Valleys and gorges—Australian mountains—Mexican mountains—Cotopaxi—Uses of mountains   30

III

SCENERY AND ITS CAUSES

Early notions as to scenery—The earth a changing scene— English landscape—The action of the sea—Coast scenery—

The power of frost, weathering, and rain — River work — Typical scenery—Scenery of the plain—The Steppes—Forest plains—Deserts—Cavern scenery   .   .   . PAGE 58

IV

THE SEA

Early notions—English love of the sea—The ocean depths— Increase of pressure with depth—Composition—Saltness— Colour—Oceanic movements—The Gulf Stream and other ocean rivers—Wave motion—Islands—The Mediterranean— Polar seas—Harvest of the sea—Uses and significance of the sea   83

V

RIVERS AND THEIR WORK

Early notions—The Euphrates—Helpful work of rivers—Rivers sometimes a menace—Various uses of rivers—English rivers— The Rhone—The Rhine—The Ohio—The Nile—Phenomena of rivers—Source—Watershed—Narrows and rapids—Water- falls : in Europe, America, and Africa—Flooding of rivers — River mouths, estuarine and deltoid—The Thames—Conditions of usefulness of rivers   113

VI

LAKES AND THEIR LESSONS

Lakes : their effect upon scenery — Promote water carriage — Affect climate — Alter the flora and fauna — Salt lakes — Story of the lochlands—Swiss lakes and lake dwellings— Harvest of the lakes—Lakes, how classified—Their formation — English lakes — The Caspian — Dead Sea — Freshwater lakes—Lake Geneva—African lakes — Curious lakes—Con- clusion   147

VII

WELLS AND SPRINGS

Introductory — Jacob's Well and Eastern springs — Origin of springs — Types of springs — Perennial — Reciprocating — Artesian — Thermal—Medical—Geysers—Salt springs—Sacred wells—Uses of springs in Nature—Conclusion   PAGE 174

VIII

THE AIR ; MAN'S VITAL ELEMENT ·

Early notions—First experiments upon the air—Pascal and Torricelli—Properties of the air—Humidity—Elasticity—Mobility—Refractive power—Cavendish and his labours—Composition of the air—Water vapour in the air—Minor components—Weather and climate—Motions, direction of winds in England—Local winds   193

IX

THE WINDS OF HEAVEN

Primitive ideas—Gods of the winds—Causes of wind—Experimental illustrations—The trade winds—Anti-trades—The winds of England—Winds as ventilators—Wind and weather charts—Ballot's weather law—Land and sea breezes—Monsoons—Cyclonic storms—The law of storms—Great hurricanes—The place of winds in Nature   211

X

THE FORCE AND THE FILIGREE OF FROST

Frost work in England—Frost force—Glaciers—Ice flowers—Frost filigree—Icebergs—Polar ice—Glaciers—Journeys towards the North Pole   230

XI

THE EARTH'S FIRES

Fire gods of antiquity—Earthquakes superficial—Great volcanic
bands—European fire region : Vesuvius—Products of eruption
—Earthquake of Lisbon—American volcanoes—The Asiatic
band—Minor centres of the igneous forces .   . PAGE 256

XII

"YE SHOWERS AND DEW"

Dew: Scriptural allusions to dew—Poetical references—Cause of
Dew—Experiments upon—Kinds of dew—General phenomena.
Rain: Rainfall — Distribution of rain — Rainless regions —
Rains of England — Rain and rivers — Rain a blessing.
Clouds: Their appearance—Difficulty of study—Why they
float—Classification—Cloud prognostics   .   .   . 279

GLOSSARY 303

INDEX   .   .   .   .   .   .   ,   . 309

LIST OF ILLUSTRATIONS

FIG.   PAGE
1. Christopher Columbus 7
2. Captain Cook 15
3. David Livingstone 16
4. Section of the Grampian Range from Inverness to Kirriemuir. (Vertical and horizontal scale the same) . 19
5. Vertical arrangement of the land . . . 20
6. Elevation of the great continents . . . 20
7a. Cavendish experiment, weighing the earth . . 23
7b. Cavendish experiment in weighing the earth . . 24
8. Lateral attraction 24
9. The earth's revolution around the sun . . . 25
10. The earth at an equinox 27
11. The earth at the summer solstice . . . 28
12. The *Santa Maria*, in which Columbus sailed across the
    Atlantic 29
13. Upper Valley of St. Gothard 34
14. Snow-Fields and Glaciers of Mont Blanc, seen from the
    top of Mont Brévent 35
15. Summit of Mont Blanc 37
16. View in the Valais below St. Maurice . . 39
17. The Matterhorn 40
18. Grand Cañon of the Colorado 44
19. The Grand Cañon, Colorado 46
20. A View on the Zig-Zag of the Blue Mountain Railway . 48
21. Ben Nevis Observatory 51
22. Gibraltar 53
23. Gorge of the Ericht, above Blairgowrie . 57

FIG.   PAGE
24. Group of beeches, Burnham   61
25. The Armed Knight and the Long Ship's Lighthouse,
    Cornish coast : illustrating bombardment by the sea   63
26. The great Báltoro Glacier, Himalayan Mountains   .   65
27. Brig o' Trams, Wick : Weathering Action   .   .   67
28. The High Woods of Jamaica   73
29. Entrance to the "Convolvulus Cave," Walsingham,
    Bermudas   75
30. The Great American Desert   77
31. Logan Stone near Land's End   81
32. Æolian Rocks, Bermuda.   84
33. Ruins of a Viking Ship   85
34. Macleod's Maidens and the Basalt Cliffs of the West of
    Skye 86
35. View on the East Coast of Scotland   .   .   .   87
36. "The Old Man of Hoy," Orkney Islands   .   .   89
37. The Land's End, showing the wasting action of the sea   91
38. The Steeple Rock, Kynance Cove, Cornwall   .   .   93
39. The Lion Rocks, Cornish Coast—a sea in which nothing
    can live   95
40. Salt pits in the west of France .   .   .   96
41. View in a Norwegian Fjord   97
42. Spring-Tides and Neap-Tides   100
43. Marine View on the West Coast of Scotland   .   .   103
44. The Stacks of Duncansby, Caithness, a wave-beaten
    coast-line   111
45. On the Nile : a Cargo Boat   .   .   .   115
46. The Wharfe : Bolton Woods   .   .   .   117
47. The Banks of the Susquehanna   121
48. View on the Congo, above Boma   125
49. The Falls of Clyde   129
50. Gavarnie Falls   130
51. Salmon Leap on the Bann   131
52. Niagara, from Goat Island   .   .   .   133
53. The Murchison Falls   135
54. "China's Sorrow" : the Yellow River in flood .   .   139
55. The Ganges at Derali   140

FIG.   PAGE

56. On the Amazon : Tropical Vegetation .   .   . 141
57. Windsor Castle from the Thames   .   .   . 143
58. On the Thames : Nuncham Courtney .   .   . 145
59. The Black Rock of Novar, Cromarty Firth   .   . 146
60. Loch Katrine and Ellen's Isle 149
61. Loch Leven   151
62. Windermere   .   .   . 159
63. Rydal Water   161
64. View up the Valais from the Lake of Geneva   163
65. Storm on Albert Lake .   .   . 165
66. The Pitch Lake .   . 167
67. Mirror Lake, Yosemite Valley .   .   . 170
68. Reflected Scenery   171
69. Terraces of Great Salt Lake, along the flanks of the
     Wahsatch Mountains, south of Salt Lake City, in-
     dicating the shrinkage of the waters .   .   . 172
70. Pool of Siloam 178
71. Section of Formation for Artesian Well .   . 179
72. St. Anne's Well, Buxton   .   .   . 181
73. Bechive Geyser, Yellowstone Park, Colorado   . 183
74. Oil Wells in Pennsylvania   .   . 185
75. The Dropping Well, Knaresborough   . 187
76. Robin Hood's Well 189
77. The Earth's garment of air   .   .   . 194
78. Wheel Barometer   .   .   . 195
79. The Atmosphere 198
80. Specimen of *Daily News* barometer chart   .   . 207
81. Specimen from *Daily Telegraph* weather chart, showing
     range of barometer 208
82. The Wind-Rose for England .   .   . 220
83. Expansive Force of Freezing Water   .   . 231
84. Glacier of Zermatt   233
85. Formation of Icebergs 234
86. Glacier of the Blümlis Alp   .   .   . 237
87. Forms of Snow Crystals (Scoresby)   .   . 238
88. Ice Crystals 239
89. Frost Flowers 241

FIG.   PAGE
90. Arctic Iceberg seen on Parry's first voyage . . 242
91. Tabular Iceberg detached from the great Antarctic Ice-barrier 243
92. The *Vega* and *Lena* moored to an Ice-floe . . 244
93. Diagram of the approximate extent of Permanent and Floating Ice around the North and the South Poles. (After Petermann) 247
94. The Antarctic Ice-wall and Icebergs . . . 248
95. Glacier with Medial and Lateral Moraines . . 249
96. The union of two Glaciers, showing junction of two Lateral into one Medial Moraine . . . 250
97. The Mer de Glace 251
98. The Palæocrystic Sea . . . 253
99. Frost-work. Luna Island, Niagara . . . 255
100. Sketch of submarine volcanic eruption (Sabrina Island) off St. Michael's, June 1811 259
101. Stromboli, viewed from the north-west . . 261
102. Vesuvius 263
103. View of Vesuvius as seen from Naples during the eruption of 1872 264
104. Lava Stream 267
105. View of portion of a Lava Stream on Vesuvius . 270
106. View of houses surrounded and partly demolished by the Lava of Vesuvius, 1872 271
107. View of the great Basalt-plain of the Snake River, Idaho, with recent cones 274
108. Mount Rainier—a Volcano, Puget Sound . 275
109. Distribution of Rain 287
110. Cloud Form—Cirrus 292
111.   Do.   Cumulus 293
112.   Do.   Stratus 294
113.   Do.   Nimbus 295
114. Cloud Scenery 296
115. Cloud Scenery—the Firth of Tay . . . 297
116. Cumulus and Strato-Cumulus 299
117. On the Lima and Oroya Railway. Among the Clouds 300

I

THE GREAT GLOBE ITSELF

"If then a picture of nature were to correspond to the require-
ments of contemplation by the senses, it ought to begin with a
delineation of our native earth.   It should depict first the
terrestrial planet as to its size and form ; its increasing density
and heat at increasing depths in its superimposed solid and liquid
strata ; the separation of sea and land, and the vital forms
animating both, developed in the cellular tissues of plants and
animals ; the atmospheric ocean with its waves and currents,
through which pierce the frost-crowned summits of our mountains."—Humboldt's *Cosmos*.

IN the far-off times of antiquity of which the earliest Intro-
records treat, the ideas of mankind concerning this ductory.
earth, the "cradle, home, and grave" of the race, were
both meagre and erroneous.[1]   In the older Biblical

[1] "The earth being an extended plain, like a table—as motionless
as that household instrument—the sun coming to take his daily peep
at it like a careful watchman on his rounds."—Milner's *Gallery of
Nature*.
    According to the poems of Homer, which are imbued with the
ideas of the ancient Greeks as to nature, and also mankind and their
ways, the earth is a great disk elevated at the edges by a lofty girdle
of mountains, round which the river ocean rolls its swelling waves.
In the centre of the disk Olympus towered up with its three rounded
summits, on which stood the mansions of the ever-happy gods, and
Jupiter, throned on its loftiest crest, looked down through the clouds

B

writings expressions are employed showing how limited was the knowledge possessed at the time they were written. Homer and the old Greeks regarded the earth as very circumscribed; they knew only the countries bordering upon the Mediterranean, and not all of them; and the people of Palestine regarded the portion of the Mediterranean which formed their western boundary as "the great" and only sea.

The Norse-men.

The hardy Norsemen, who long afterwards bore so important a part in the making of England, had ideas (which are not lacking in interest to-day) as to the origin and nature of the earth. In this "childhood of the world" their teachers, the Scalds, in their habit of personification, spoke, in the Edda, their sacred book, of the rivers as "the blood of the valleys," or as "the sweat of the earth"; and of the world itself as "the vessel that floats on the ages," and as "the daughter of the night." Another account from the same source described the earth as formed from the body of Ymir; from his blood the seas and waters; from his flesh the land; from his bones the mountains; from his jaws the stones and pebbles; from his hair the trees. The Norsemen loved the sea, and endeavoured to fathom its mysteries; they regarded it as encircling the land like a ring; and they spoke of four dwarfs that guarded the cardinal points. They had also a glimmering of a controlling deity, and of a supernatural world.

Teachings of the ancient schools.

Before this time several of the great teachers of the

and saw the restless crowd busy at his feet. The land divided into two parts by the blue sheet of the Mediterranean, stretching far away to the very verge of the disk, like the raised figures which ornament the front of a shield. Down from the heights of Olympus the immortals contemplated in one glance all the peninsula of Greece, the white isles of the Archipelago, the coasts of Asia Minor, the plains of Egypt, the mountains of Sicily, inhabited by the Cyclops, and the pillars of Hercules—the boundary of the ancient world. All round, above the track inhabited by man, stretched the crystal dome of the firmament, borne up by the two columns of Atlas and Caucasus.— *Reclus.*

ancient schools of philosophy had some perception of
a wider and truer view of nature than that which
commonly obtained. In opposition to the notion that
the earth was immovable and its features permanent—
that its mountains, plains, rivers, lakes, and seas were
for ever fixed—the Pythagoreans had knowledge of the
conversion of the sea into land, and of land into sea
again, as explaining the occurrence of marine shells in
inland rocks; of the gradual reduction of hills to
plains; of the scooping out of valleys by floods, and the
carriage of the waste to the sea; and of the gradual
alteration in the earth's surface by the formation of
new land.

Aristotle, whose teaching was directed in part to the
observation of the aspects of nature, and who has been
pictorially represented in the midst of his pupils point-
ing to the earth as a source of inspiration, was impressed with the beauty and mystery of the phenomena
of nature. In a passage that may illustrate the
"golden flow" of eloquence of this ancient teacher it is
remarked: "If there were beings who lived in the
depths of the earth in dwellings adorned with statues
and paintings, and everything which is possessed in
rich abundance by those whom we esteem fortunate;
and if those beings could receive tidings of the power
and might of the gods, and could then emerge from
their hidden dwellings through the open fissures of the
earth which we inhabit; if they could suddenly behold
the earth and the sea, and the vault of heaven; could
recognise the expanse of the cloudy firmament, and the
might of the winds of heaven, and admire the sun in its
majesty, beauty, and radiant effulgence; and, lastly,
when night veiled the earth in darkness, they could
behold the starry heavens, the changing moon, and the
stars rising and setting in their unvarying course
ordained from eternity, they would exclaim: 'There are
gods, and such great things must be the work of their
hands.'" The native land of Aristotle was singularly

Margin notes: Aristotle's eloquence. — Aristotle's teachings.

beautiful, and would strongly appeal to his imagination.[1]

In the early centuries the earth, with its mountain ranges, its expansive seas dotted with islands, its rivers and waterfalls enlivening the land, its mysterious caverns and deep places, and with its aspect varying under the play of the great physical forces, was regarded as an extended plain. In the Homeric poems it is described as environed by the dark river Oceanus with unknown lands beyond. Of the people regarded as mystical were the Cimmerians, the dwellers in darkness beyond the boundaries of the Black Sea; the Hyperboreans, beyond Boreas or the extreme north ; the Ethiopians (the sun-burnt race this term betokens), occupying the south of Egypt; and the Amazons, beyond the Carpathian Mountains. They also spoke of a desirable country, Elysium, and of Colchis, the land of magic. The heavens were regarded as a brazen dome, supported by huge pillars, the sun rising from a lake in the east, traversing the sky in a chariot driven by Phœbus, the sun-god, retiring for the night into the ocean, and again returning to the east to renew his daily course. But this was not the unanimous teaching of the times ; others considered the earth as resembling a cylinder, a boat, or a square with a margin of mountains.

From 900 B.C. to the time of Herodotus, in the fifth century A.D., knowledge of topography was extended by the addition of Sicily, Corsica, Sardinia, South Gaul, Spain, and the Adriatic Sea, the Mediterranean area being the centre of civilisation at this period. Soon afterwards North Africa, Asia Minor, and the coasts of the Black Sea became known. Ptolemy, in the second century of the Christian era, also enlarged the geographical knowledge of his day. Thales the philosopher

[1] Grecian scenery presents the peculiar charm of an intimate association of land and sea, of shores adorned with vegetation, or picturesquely girt round by rocks gleaming in the light of aerial tints, and of an ocean beautiful in the play of the ever-changing brightness of its deep-toned waves.—*Humboldt.*

(who made the foundation experiment in electrical science), in the seventh century B.C., first taught the globular shape of the earth, and this new idea was strengthened by subsequent discoveries. Pythias, of Marseilles, by his enterprising voyages revealed the coasts of France and Spain more fully; and, passing through the Straits of Gibraltar into the Atlantic, discovered the English land. Aristotle, the teacher of Alexander the Great, believed in the spherical form of the earth; at this time mathematical knowledge being brought to bear upon the subject.

The growth of our knowledge of the physical condition of the earth has been a long and laborious process, the work of many men in many lands. The astronomer, by nightly watch and elaborate computation, has shown the great globe we inhabit to be but one of a group of worlds circling round the sun as the foundation; the exploring traveller has made known the most distant parts of the earth; and the navigator has steered a bold course across the seas. Land and water have so yielded their secrets that the world's map is now almost complete. The geologist has penetrated below the surface to read the records of the rocks, and has brought to light evidence of worlds preceding the present one, each with special forms of animal and vegetable life, and a changed configuration of surface, whilst the physicist has experimented upon its mass, its atmospheric cloak, and upon the natural forces that govern it, until the great physical facts of our earthly home are largely revealed.

In the extension of geographical knowledge, commerce has played an important part. When the primitive people who were hunters, and afterwards keepers of cattle, or cultivators of the soil, were succeeded by races of a higher style of life, who began to exchange their commodities with the peoples around them, geographical information spread apace. We read of the Midianites carrying the spices of their country as merchandise, and of the Phœnicians sailing their ships

(margin note) Earth knowledge of various kinds.

(margin note) Geographical knowledge enlarged by commerce and military enterprise.

along the Mediterranean coasts from east to west, passing through the Straits of Gibraltar into the open Atlantic, and even touching these islands of ours. Commercial ports were established at convenient centres along the Mediterranean seaboard; and as navigation advanced, longer voyages were undertaken in better-appointed vessels; whilst the handling of imports led to the study of the lands from whence they came.

Knowledge of the earth's surface has also been increased by military enterprise, as the expedition of Alexander the Great eastward, and by the establishment of the Roman Empire. Sicily was added some 250 years before the Christian era; then Sardinia and Corsica; and northwards, Germany, France, and the British Isles were subsequently made known.

For many centuries but little progress was made in exploring the great oceans. In the fifteenth century, however, there was a remarkable intellectual awakening in the countries of Western Europe, leading to energetic action. Amongst the foremost personages of that time the figure of Christopher Columbus [1] is conspicuous. This eminent man, who was in possession of the most complete scientific knowledge of his day, had also tact, courage, and patience. His belief in an undiscovered world far away in the west was no visionary idea unsupported by evidence, but was a conviction based upon

[1] "The Personality of Columbus."—*The Times*, 8th August 1892.

"What manner of man was this Christopher Columbus? The cold stone of his sculptured features and two or three old and faded portraits are the only means we have of judging of his personality. Comparing the pictures existing in Genoa, the following general characteristics may be noted :—

"As a youth, ruddy-faced, with auburn hair, such a one as we rarely meet nowadays amongst the sallow-skinned, dark-eyed boys of Northern Italy.

"He was evidently a man of massive frame and powerful build, his bearing majestic and dignified ; the facial expression is one of great gravity, amounting almost to sternness. The face is oval, the forehead high and ample, and furrowed deep by the incessant working of an active and energetic mind ; the clear azure eyes have a piercing gaze, whilst the contraction of the eyebrows produces almost a frown " (Fig. 1).

experience, computation, and close observation.   The

FIG. 1.—Christopher Columbus.

great ocean river, the Gulf Stream, which flows from
Florida diagonally across the Atlantic to the shores of

Columbus
and his
great
voyage of
discovery.
Europe, had brought from time to time to those outlying western islands of the Old World—Madeira and
the Azores—evidences of an unknown land beyond the
wide western waters. Branches of strange trees and
rudely-carved pieces of wood, and upon one occasion
the bodies of two Indians, were washed ashore. These
were full of suggestion to the active mind of Columbus,
and, with other evidence, induced him to project the
nautical enterprise by which the western hemisphere,
with its wealth of natural resource, was opened up.

Ships of
Columbus.
After difficulties and discouragements that would
have disheartened ordinary men, and after weary years
of waiting (his scheme having been theoretically complete for nearly twenty years), this remarkable man was
able to realise his life's desire.[1] The vessels for the
expedition were but ill prepared for so great a voyage.
Only one of the three was full-decked, the other two
being light barques or caravels, not superior to the
coasting craft of to-day. From the *Santa Maria*
Columbus floated his flag (Fig. 12). The *Pinta* was
commanded by Pinzon, a captain of influence in the
country; and a third vessel, with simple lateen sails,
was the *Nina*. The combined crews numbered 220
men. Thus equipped, Columbus, in his fifty-sixth
year, set sail in the month of August 1492. The
journey to the Canary Islands was plain sailing, but
from this point the voyage was over unknown waters.
How ably the commander carried into execution his
great design has been well told by his biographer : " He
regulated everything by his sole authority; he super-
intended the execution of every order, and allowing
himself only a few hours for sleep, he was at all other
times upon deck. As his course lay through seas which

[1] Columbus is said to have heard in his dreams a voice saying :
"God will cause thy name to be wonderfully resounded through the
earth, and give thee the keys of the gates of the ocean, which are
closed with strong chains." This prediction has been fulfilled, and
the fourth centenary of the discovery of America has been observed
with enthusiasm in America, at Madrid, and in England.

had not formerly been visited, the sounding-line or
instruments for observation were continually in his
hands.   After the example of the Portuguese discoverers, he attended to the motion of tides and
currents, watched the flight of birds, the appearance of
fishes, of seaweeds, and of everything that floated on
the waves, and entered every occurrence with a minute
exactness in the journal which he kept." [1]

Steering a westerly course, he soon encountered the
trade winds, which bore him onward with steady
regularity.   When a hundred leagues from the Canaries
he found that the compass was no longer true to the
pole, and its variation gave him grave concern.   When a
hundred leagues farther out at sea his ships became
entangled in the immense marine meadow known as
the Sargasso Sea, an area of the Atlantic larger than
the whole of France, covered with floating seaweed.
This alarmed the sailors, who thought they had touched
the end of the world, and that the vessels would be
stranded upon its outer edge.   The captain, however,
pursued his westerly course, but for long with so few
encouraging prognostics that the mariners grew mutinous.   Columbus, not less skilful in the control of men
than in the management of his ships, endeavoured to
work upon their ambition, their avarice, or their fears.

Discovery of the trade winds and of the variation of the compass.

Mutinous crew.

[1] Columbus discovered the trade winds ; encountered the Sargasso
Sea, an immense area of the ocean covered with seaweeds and marine
social plants ; observed the variation of the compass needle, and suggested that latitude could be inferred from the amount of dip, and
longitude from the declination ; but he trusted most to astronomical
observation, saying that, "He who understands it may rest satisfied,
for that which yields it is like unto a prophetic vision."

Queen Isabella, who may be regarded as the founder of the expedition, showed her appreciation of scientific knowledge.   Writing to
Columbus, September 1493, that although he had shown in his under-
takings that he knew more than any other living man, she counselled
him nevertheless to "take with him Fray Antonio de Marebend, as
being a learned skilful astronomer."

In this way the Queen indicated the necessity for combining
advanced theoretical knowledge with skill in practical seamanship.

But the commander's arts were at length nearly
exhausted. Officers and men "assembled tumultuously
on the deck, expostulated with him, mingled threats
with their expostulations, and required him instantly
to tack and return to Europe.[1] In this extremity
Columbus promised that if land were not sighted within
three days he certainly would return. Notwithstanding
the excitement of the situation this short period was
coolly calculated. The sounding-line had touched the
bottom, and indicated shallowing for shore; flocks of
land birds as well as sea-fowl were observed; branches
of trees floated past,—which, with the rapid veering of the
wind, announced that land was near. But there was no
time to spare, and the fate of the expedition seemed to
tremble in the balance. It was towards the close of the
last day, according to Robertson's interesting narrative,
when occurred an event which was a turning-point in
the world's history, and which might well form the
subject of a commemorative fresco for a building in
which the Old World and the New are jointly interested.
"About two hours before midnight Columbus, standing
on the forecastle, observed a light at a distance, and
privately pointed it out to Pedro Guttierez, a page
of Queen Isabella of Spain, whose important share in
the discovery of America has been referred to.
Guttierez perceived it, and, calling to Salcedo, comp-
troller of the fleet, all three saw it in motion as if it
were carried from place to place. A little after midnight the joyful sound of 'Land! land!' was heard
from the *Pinta*, which kept always ahead of the other
ships. But, having been so often deceived by fallacious
appearances, every man was now become slow of belief,
and waited in all the anguish of uncertainty and im-
patience for the return of day. As soon as morning

The first sight of America.

[1] " Through the broad waste of waters drear and dark,
'Mid wrathful skies, and howling winds and worse,
The prayer, the taunt, the threat, the muttered curse,
Of all his brethren in that fragile bark."—*Tupper.*

dawned, all doubts and fears were dispelled. From every ship an island was seen about two leagues to the north, whose flat and verdant fields, well stored with wood and watered with many rivulets, presented the appearance of a delightful country." Thus was won the first outpost of the western world by the skill and daring of a great navigator; and soon afterwards geographical knowledge received a more important accession by the discovery of the American mainland. **A delightful land.**

Another notable maritime discoverer of this enterprising age was Vasco de Gama, a native of Portugal, who was the first to double the Cape of Good Hope and reach India by sea. This voyage forms the subject of the Portuguese national poem—"The Epic of Commerce," it has been called, as it prepared the way for an extensive trade with the rich countries of the East. No navigator had rougher seas to contend with than De Gama. The waters of the "roaring forties"[1] (the parts of the South Atlantic between the 40th and the 50th parallel south latitude) are proverbial for their tempestuous character, the navigation not being without danger even to the powerful steam vessels of the present day. The skill and determination of this navigator overcame all difficulties. A contemporary writer says of him—" He besought all not to take account of their labours, since for that purpose they had ventured upon them; and that they should put their trust in the Lord that they would double the Cape." Of his behaviour at a very critical stage of the journey it is further affirmed "he gave them great encouragement, always taking part with them in hardship, coming up at the boatswain's pipe as they all did. So they went on **Vasco de Gama as a discoverer.**

---

[1] Diaz, returning from his voyage in 1487, discovered the Cape. At first it was called Cabo dos Tormentos, or the Cape of Storms, afterwards changed to the Cape of Good Hope.

> " At Lisbon's Court they told that dread escape,
>   And from the raging tempests named the Cape ;
>   Thou southmost point, the joyful King exclaimed,
>   Cape of Good Hope be thou for ever named."

standing out to sea till they found it all broken up
with storm, with enormous waves and darkness; and as
the days were very short, it seemed always night. The
masts and shrouds were stayed, because with the fury
of the sea the ships seemed every moment to be going

Great troubles at sea. to pieces. The crew grew sick with fear and hardship,
because they could not prepare their food, and all
clamoured for putting back to Portugal, saying that
they did not choose to die like stupid people who
sought death with their own hands; thus they made
clamour and lamentation." When persuasion failed
the captain "comported himself very angrily, swearing
that if they did not double the Cape he would stand

The Cape of Good Hope doubled. out to sea again as many times as necessary until the
Cape was doubled." The heroic commander did not
vainly threaten, but "sailed larger," so as to round the
Cape; and as they did not, steering an easterly course,
find land, "they knew they had doubled the Cape."

Magellan circumnavigates the globe. The western voyage of Columbus and the eastern
voyage of De Gama were surpassed by the achieve-
ments of Magellan,[1] who first circumnavigated this
great globe. This famous sea-captain, who was of
Portuguese birth, had control of five small vessels of
from 75 to 120 tons. They were small and unsea-

His poor equipment. worthy, "ill ordained to sail to the Canaries," testifies
a writer of the times, and with "knees of touchwood."
The ships carried, as cargo for exchange and trading
purposes, copper, quicksilver, coloured cloth, and silk
jackets. Like Columbus, Magellan had on shore successfully contended with intrigues and jealousies; and once
fairly at sea, the greatest hardihood and determination
were shown. Starting from the port of Seville in the
year 1519, the captain pursued a south-westerly course.
The "Milton of Portugal," Camoens, who describes in

---

[1] *The First Voyage Round the World.* Notes by Lord Stanley of
Alderley. Hakluyt Society, 1874.
"Magellan's enterprise was the greatest ever undertaken by any
navigator." His vessels were the *Conception*, 90 tons; *Victoria*, 85;
S. Antonio, 120; *S. Trinity*, 110; and *Santiago*, 75 tons.

detail the voyage of De Gama, in the *Lusiad*, the Epic of Commerce, has a word of praise for Magellan.[1] ·

Magellan successfully crossed the Atlantic from north-east to south-west; passed through the straits which afterwards bore his name; and, for the first time, sailed right across the broad Pacific. The enormous extent of previously-unknown seas traversed by Magellan may be realised to some extent by a few extracts from the ship's log-book. "We emerged from the straits (which are about 400 miles long) 27th November 1520, and sailed between west and north-west 9858 miles to the equinoctial line; then west and north-west 2016 miles to the Ladrones, and then 1000 miles to the Moluccas." At the next group of islands—the Philippines—Magellan was killed in a skirmish with the natives. The expedition, although thus losing its "mirror, light, and true guide," was continued, and the *Victoria*, Magellan's flagship, was brought to the harbour from which it set forth, having sailed completely around the world. The true magnitude of the earth, which Columbus had under-estimated, was established, and the foundation laid for subsequent discovery by sea and land. A glimpse of the character of Magellan may be obtained from his autobiography and the records of his contemporaries. At the outset of this voyage he thus addressed his crew: "I command you on the part of the Sovereign; and on my part beseech you and charge you, that with respect to all that you think is fitting for our voyage, both as to going forward and as to turning back, you give me your opinions in writing, each one for himself declaring the circumstances and reasons why we ought to go forward or turn back, not having respect to anything for which you should omit

In the great Pacific Ocean.

His instruction to his crew.

[1] " Along these regions from the burning zone,
    To deepest south he dared the course unknown,
    While in the Kingdom of the rising day,
    To rival thee [1] he holds the western way."

    1 De Gama.

to tell the truth." The captain's reasons were all for going forward, "and he swore by the habit of St. James which he wore, that so it seemed to him to be for the good of the fleet." Magellan lost two of his ships in passing through the straits which now bear his **Diffi-** name. During one part of the voyage over three **culties and** months elapsed without sighting land. The provisions **dangers of** **the voyage.** were exhausted, and the sailors were reduced to the extremity of eating even old leather on board and drinking foul water. At the Philippines, besides losing its commander, the small fleet lost one of its ships by fire ; but the *Victoria*, with a party of Indians on board and forty-six Spaniards, crossed the Indian Ocean, doubled the Cape, reversing the course of De Gama, and, after a stormy voyage, reached the shores of Spain, the high achievement being acknowledged by public thanksgiving in the churches. The great oceans were thus crossed if not explored ; and with the increase in mental activity that now characterised the more promi- nent European nations, knowledge of the earth began to grow apace. In the promotion of geographical dis- **More re-** covery England now began to bear part. In the **cent geo-** eighteenth century Captain Cook (Fig. 2) circumnavigated **graphical** **discovery.** the southern hemisphere, discovering New Zealand and thoroughly opening up the South Pacific. In the present century Ross explored the coast of North America ; whilst Humboldt, one of the most scientific of travellers, examined the countries of South America ; and Sir Stamford Raffles and Sir James Brooke further revealed the geography of the East Indies.

Previous to the year 1840 but little was known of Australia beyond its borders ; the interior was a blank upon the map. In 1841, however, Mr. Eyre crossed from Fowler Bay, on the south coast, to St. George's Sound, passing over a thousand miles of new ground. In 1844 Captain Sturt travelled from the Darling River **Australian** northwards to the centre of the continent; but failed **explorers.** to accomplish his design of reaching the Gulf of Car-

pentaria, on the northern seaboard, owing to the want
of food and water which the barren country would not

Fig. 2.—Captain Cook.

yield.   In the same year Dr. Leichardt, proceeding
from the same district by the coast, touched this gulf,

and from thence travelled to Port Essington on the
western coast, a distance of not less than 1800 miles,
over previously untravelled land.   The labours of
Livingstone (Fig. 3), Stanley, and others have let in

Fɪɢ. 3.—David Livingstone.

the light upon "The Dark Continent," so that recent
maps show few unexamined tracts of either land or
water.

The land
surface.

The land is man's especial domain.   It may be
thought of as spreading horizontally as shown in
ordinary maps, or reaching vertically as indicated in
contour-line maps.   The total area is not far short of

200 million square miles,[1] of which the land occupies
52 millions. The great land masses are as irregular
in shape as in the manner of their distribution.
The land is chiefly in the northern hemisphere in the
ratio of 38 to 14, and in the eastern as compared with
the western, where the same proportion holds. If the
earth be thought of as having England in the centre
of one hemisphere and New Zealand the centre of the
opposite one, then 49 parts out of 52 of the land
would be in the London half, a fact of immense im- England in
portance in relation to our commercial position. the centre
England is thus not situated at the verge of the land, of the land.
as the ancients supposed ; but is centrally placed with
reference to the nations of the earth. The Atlantic
is now crossed by our greyhound liners in a few days,
and our mercantile intercourse with America is of the
first importance. The European ports are easily
reached ; our customers are all around.

The configuration of the land largely determines the Articula-
capability of a country for progress in civilisation. tion and
Europe and North America are centrally placed be- progress.
tween the zones of extreme heat and extreme cold ;
their temperature is favourable to human activity.
Each of them has independent ports and distinctive
physical regions ; each is well articulated, the land

[1] The earth's surface in millions of square miles : total area, 197 ;
water, 145 ; land, 52. The eastern continent is found to contain 36
million square miles, or about a fifth of the whole surface of the
globe. Its greatest extension is from the Cape of Good Hope to the
north-east of Asia, 10,000 miles. The western continent contains
14 million square miles of surface, and upon it a line of 8000 miles
could be drawn between the most extreme points. The eastern
hemisphere is conventionally divided from the western at the 20th
meridian west longitude and 160° east.
The Pacific Ocean occupies more than two-fifths of the globe, or
80 million square miles. It is 11,000 miles from north to south, and
10,000 from east to west.
The Atlantic area is 25 million square miles ; 10,000 miles from
north to south, and averaging 2000 miles across.
The Indian Ocean contains 13 million square miles, and the Mediterranean Sea a million square miles. The sea = $2\frac{1}{2}$ of the land area.

C

stretching seaward, forming peninsulas, and the seas deeply indent the land. Each has great length of coast with natural harbours favouring commerce; and in Europe, with the exception of Russia, no part is far away from the sea-board. There are other physical advantages giving these continents the premier place.[1] Europe, it has been said, has giant limbs without a body; Asia has an immense body, and limbs equally immense. Its central portions are of but little importance; but the countries in contact with the sea, as India, China, and Japan, are well populated and progressive. The most backward continents are those whose physical position is unfavourable, as Australia, whose centre is still largely unoccupied and unknown; and Africa, whose interior has only recently been discovered.

Africa wanting in articulation.

The Polar regions are still unexplored. The North Polar Ocean as connecting important countries has been frequently attempted, but the farthest point yet attained was by Captain Markham, 83° 20′ north, or some 400 miles from the Pole. A new expedition is being prepared by Dr. Nansen, who, having crossed over Greenland, is desirous of extending our knowledge of the Arctic Ocean. His ship, specially built for the purpose, is provided with steam power and is equipped with the most approved appliances for the voyage, and with provisions for several years. Observation shows that a constant and powerful current passes from the north near the eastern sea-board of Greenland, and the theory of Dr. Nansen is that a corresponding counter current proceeding to the Pole passes near the New Siberian Islands—a conclusion founded upon several discoveries—and the intention is to drift towards the Pole with this current and back again with the Greenland or some other south-running current. The

Arctic exploration.

---

[1] But not only the great continents as a whole; it has been remarked that "every mountain, every headland, every islet, every lake, river, or rivulet plays its part in the history of mankind."

Kirriemuir.

Cat Law.

Glas Meal.

Dee.

Ben Macdhui.

Spey.

Inverness.

FIG. 4.—Section of the Grampian Range from Inverness to Kirriemuir. (Vertical and horizontal scale the same.)

equipment includes special boats, which may be used for the return journey should the vessel be lost in the ice. The best scientific instruments will be employed with the object of working out interesting problems depending upon an extension of our knowledge of Polar conditions.

Of the physical characteristics of the land, its elevation above the sea-line deserves distinct mention in its importance to mankind. A change of climate is caused by the difference of a few hundred feet in the vertical direction that some hundreds of miles, travel on the horizontal would hardly equal. Sometimes the land (as parts of Holland, and portions of the Lombardy plain) lies below the sea-level, and has to be protected by dykes, often at great expense, from encroachments of the waters. In maps showing the elevation the height must be exaggerated to be instructive. An illustration to true scale is shown at Fig. 4.

Some regions are level with the sea, as the delta of the Ganges (Fig. 5) and other rivers; such are in continual danger of inundation. Then there are low-lying countries, in which our own may be included (the majority of the people living at a height of a few hundred feet), possessing navigable rivers, and where canals,

How elevation determines climate.

railways, and roads are easily constructed. On account of their superior internal communication these countries are favourably situated for manufacturing development. Then there are high plains and plateaus, as parts of Spain and South America, where locomotion is . difficult, and communication with neighbouring countries, or the coast, is carried on at great disadvantage.[1]

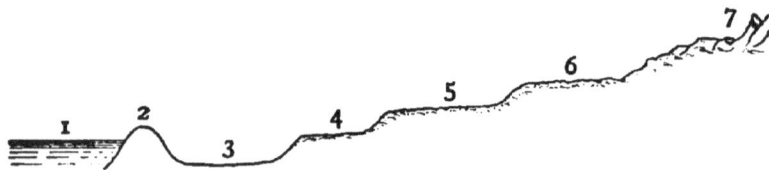

FIG. 5.

1. The sea-level, the measuring base.
2. A dyke or embankment for keeping back the waters.
3. Polder or reclaimed land below the sea-line.

4. Deltoid land slightly above the sea-level, exposed to flooding.
5. Low plains, the locality of most large cities.
6. Plateaus or high plains.
7. Mountain regions.

[1] The most recent estimate of the elevation of the great continents, supplied by Mr. Eli Sowerbutts, of the Manchester Geographical Society, is :—

	Mean Elevation.
ASIA	3000 feet.
AFRICA	2000 ,,
SOUTH AMERICA	2000 ,,
NORTH AMERICA	1900 ,,
EUROPE	940 ,,
AUSTRALIA	800 ,,

FIG. 6.

An examination of the mountain heights and the deep places of the earth has made us acquainted with the structure of "the crust," as it is called; for although this globe is now thought to be nearly solid, it may be convenient to adopt the old form of expression. This outermost layer of the great globe we inhabit contains a record of its past history, bearing witness to the play of physical forces in bygone times, and revealing the style of worlds that existed before the present order of things. The wash of the sea, the action of the rivers, the force of the fires which are contained within the earth, are all recorded in the rocks. In the Peak of Derbyshire, and amongst the mountains of Wales, for example, may be traced the cooled lava that was once shot up hot from below.[1] The English land is largely the work of the waters of the seas of past ages. The sandstone tracts were the shores of ancient seas; the slates are the

The earth's crust a record of past ages.

Destroying and renewing forces.

The general level of all lands is about 2000 feet. Further exploration of the plateau region of Africa will probably reveal a still greater elevation.

An empirical rule for the range of visibility of the earth's surface from a given height :—

$$\sqrt{\text{Height in feet} + \tfrac{1}{2}\text{ height in feet}} = \begin{cases} \text{number of miles visible} \\ \text{from the sea-level.} \end{cases}$$

Thus if the elevation of Mont Blanc = 15,720 feet,
Its distance visible from the sea-level =

$$\sqrt{15720 + \frac{15720}{2}} = 153 \text{ miles.}$$

[1] The increase of temperature with depth averages 1° Fahr. for 60 feet. At the bottom of a coal shaft near Manchester, 2151 feet deep, was recorded a temperature of 75° Fahr., the temperature at the surface being 51° Fahr. Water from an artesian well at Grenelle, near Paris, 1800 feet deep, registered 81° when the surface temperature was 60° Fahr. The hot springs at Bath and Buxton also indicate the heated condition of the inner crust of the earth in our own country, and the Geysers of Iceland, bursting forth in a boiling state, afford still more marked evidence of the earth's interior heat. The Breidden Hills, near to which the British chieftain Caractacus made his last stand against the Roman arms, are principally composed of igneous rock—are, in fact, extinct volcanoes.

hardened mud of vanished rivers, and limestone and
chalk were formed by the accumulation of the shells
and shields of countless small marine creatures, as are
the coral reefs and islands on the ocean floors of our
time.   In these primitive ages the earth presented
geographical features unlike those which now prevail.
Old ocean beds have been raised to the dignity of land,
and fitted for the abode of man by the operation of
great physical forces acting for long periods, and great
tracts of prehistoric land are now beneath the waters,
receiving daily deposits from the sea and rivers, including animal and vegetable remains, which, with
evidences of the handiwork of man, are being sealed up
to tell to distant ages the story of to-day.[1]

The prob-
lem of
Cavendish.
Whilst knowledge of the earth's surface and crust
has been made known, inquiries as to the physical forces
governing our planet as a whole have been advanced.
Amongst the scientific workers who have extended the
boundaries of knowledge in this direction may be mentioned the Honourable Henry Cavendish, a connection
of the ducal house of Devonshire, who, endowed with
ample fortune, devoted his life to scientific research and
experiment with the greatest success.

Cavendish has been described as "the man who
weighed the world."   This tremendous problem was
solved by comparatively simple experiments, the arrange-

[1] Mr. Ruskin thus describes the process of rock forming : " If
one of those little flakes of mean sand, hurried in tremulous spangling
along the bottom of the ancient river, too light to sink, too faint to
float, almost too small for sight, could have had a mind given to it as
it was at last borne with its kindred dust into the abysses of the
stream, and laid (would it not have thought) for a hopeless eternity
in the dark ooze the most despised, forgotten, and feeble of all earth's
atoms ; incapable of any change ; not fit down there in the diluvial
darkness so much as to help an earth-wasp to build its nest or feed
the first fibre of a lichen ; what would it have thought had it been
told that one day, knitted into strength as of imperishable iron,
resisting the air, infusible by the flame, out of the substance of it,
with its fellows, the axe of God should hew that Alpine tower, that
against it, poor helpless mica flake, the wild north wind should rage
in vain ; beneath it—low fallen mica flake—the snowy hills should

ments for performing which were shown at the Loan
Exhibition at South Kensington, with other original
apparatus of great interest, some years ago.   The experiment depends upon the
fact that the force we describe
as gravitation is universal in
the direction of its action,
and not simply vertical, as at
first sight it appears to be.
How bodies attract each other
in the horizontal direction
was shown at the exhibition
just referred to by an instrument whose index moved
under the attractive pull of a person approaching it,
a heavy man causing a marked deflection, a lighter
weight a movement of the index proportionately less.

Gravita-
tion uni
versal.

FIG. 7a.

In the famous Cavendish experiment a light beam of
wood, carrying at each end of it a ball of lead of about
2 inches in diameter, was carefully suspended by a thin
wire free from torsion.   Two large leaden balls were by
means of a turning table made to approach the smaller
ones on opposite sides, as in the diagram Fig. 7a, in
which the machinery for placing them in position is
omitted.   As attraction is a universal force acting in

he bowed like flocks of sheep, and the kingdoms of the earth fade
away in unregarded blue, and around it, tired wave-drifted mica flake,
the great war of the firmament should burst in thunder, and yet stir
it not, and the fiery arrows and angry meteors of the night fall blunted
back from it into the air, and all the stars in the clear heaven should
light one by one as they arose new cressets upon the points of snow
that fringed its abiding place on the imperishable spire ? "

"The waters wear the stones : thou washest away the things which
grow out of the dust of the earth ; and thou destroyest the hope of
man."—JOB. xiv. 19.

> "The hills are shadows, and they flow
>   From form to form, and nothing stands ;
>   They melt like mists, the solid lands,
> Like clouds they shape themselves and go."
> *Tennyson,*

every direction, it follows that the small leaden balls would be drawn by the heavier ones by what is called a "couple" in mechanics. The movement is very slight, and has to be watched through a telescope. The large balls are now reversed in position, as shown in Fig. 7b, when the torsion movement will be in the opposite direction. The total amount of motion gave the basis

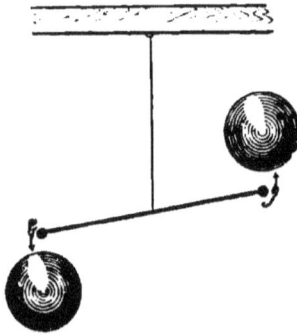

for an elaborate mathematical expression, from which the weight of the earth was proved to be over five times that of the same bulk of water. Vernon Boys has repeated this experiment, using quartz fibre, and the immense globes of lead which Cavendish used are not now necessary.

FIG. 7b.

Besides being weighed by Cavendish, this earth has been tested as to its specific gravity by other experiments. One of them was to ascertain how far a great mountain mass—Schichallion in Perthshire was chosen for the purpose

Lateral gravitation.

FIG. 8.

—attracts a plummet from the true perpendicular by reason of its local action. The verticality of the plumb line is interfered with by the presence of the mountain mass (Fig. 8), and the measure of this departure from the perpendicular towards the mountain enables the mathematician to deduce the weight of the earth.[1]

As regards its planetary relations, the earth has to be thought of, difficult as is the conception, as a mighty globe nearly 8000 miles in diameter,[2] and heavier than if composed of solid stone.

An introductory account of the earth, viewed as a whole, should take note of its rapid, if unperceived, motions. Galileo, the eminent Florentine nobleman who employed his time and means in scientific research, demonstrated its rotation by letting fall from the leaning tower of Pisa bullets of lead which were always found to be slightly deflected towards the east. The earth rotating towards the east, the top of the tower moved in a larger circle than the base, and the bullets would, on being taken to the summit of the tower, acquire a higher eastward velocity, which would, under the principle of inertia, be maintained, and so, in falling, would, as it were, overrun the surface of earth in the direction of its rotation. Similar experiments were repeated under more precise conditions by Sir Isaac Newton; and, subsequently, at Bologna, a tower 300 feet high was employed, the night-time being chosen, when the vibration of the tower was diminished. The balls let fall alighted upon a cake of wax, and it was found that in every case there was an eastward deviation. At the beginning of the present century, at Hamburg, a loftier tower was used, and balls varying from one to six inches in diameter, and of different metals, were

The earth rotates.

---

[1] Estimates of the earth's specific gravity by various methods :—

1. Deflection of plumb line by Schiehallion, Perthshire .	4·713
2. Pendulum vibrations on mountains . . .	4·857
3. Torsion balance of Cavendish . . .	5·480
4. Value from Vernon Boys' experiments . . .	5·527

[2] Equatorial diameter . 7,925·6 miles.
Polar diameter . . 7,899·2 ,,
Difference . . . 26·4 ,,
Mean circumference . 24,857·5 ,,
Surface . . . 197 million square miles.
Cubic contents . . 256,000 ,, cubic ,,
Weight . . . 5,852 trillion tons.

employed. Their descent corroborated the conclusions previously arrived at.[1]

The earth's revolution around the sun. The revolution of this great globe around the sun as a centre, and causing the seasons, as indicated in the illustrations (Figs. 9, 10, and 11), is not so easily proved. There is no easy demonstration of it; but all celestial appearances and phenomena indicate it, and there is no

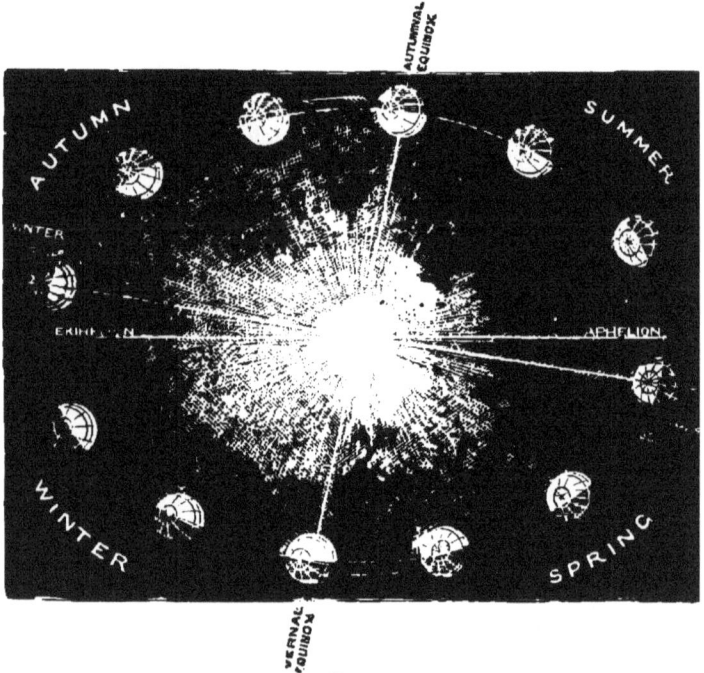

FIG. 9.

rebutting evidence. Such motion also accords with that of the earth's fellow planets which can be observed describing circles around the sun.

[1] It is not surprising the ancients did not realise the rotatory motion of the earth, which is concealed by its perfect regularity. The earth

"spinning sleeps
On her soft axle; while she paces even
And bears thee soft with the smooth air along."
Paradise Lost, Bk. viii.

Whilst the larger motions of the earth and its celestial relationships are well known, its inner movements are also under close inspection. "In each of the magnetic observatories," remarks Clerk-Maxwell, "throughout the world an arrangement is at work by means of which a suspended magnet directs a ray of

This globe a centre of physical forces.

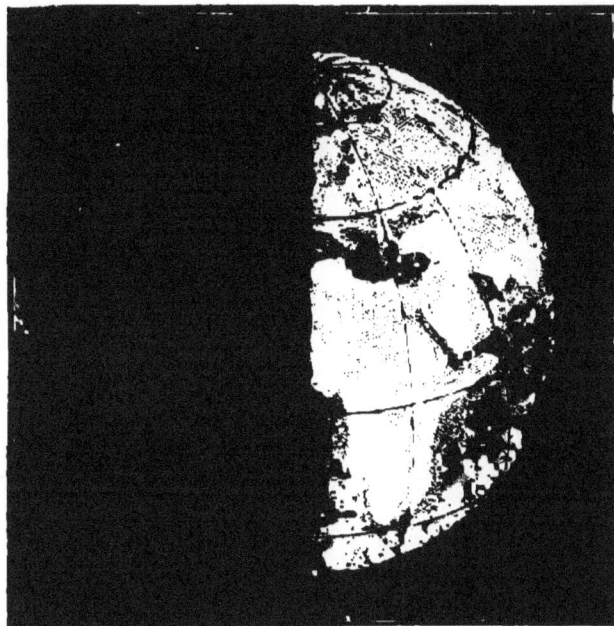

Fig. 10.—The earth at an equinox.

light on a prepared sheet of paper moved by clockwork. On that paper the never-resting heart of the earth is now tracing in telegraphic symbols, which will one day be interpreted, a record of its pulsations and its flutterings, as well as that slow but mighty working which warns us that we must not suppose that the inner history of our planet is ended."

As to its origin and ultimate destiny, a recent writer remarks : "The history of the earth is strictly analogous to that of man; both have a nebulous or chaotic state, a fiery or sunlike infancy and youth; a middle age in which all vital phenomena are fully manifested, and an old age with decreasing energy and ultimately death."

Nature's workings uniform. "The present order of things," says Professor Boyd Dawkins, "has been proved to have sprung out of

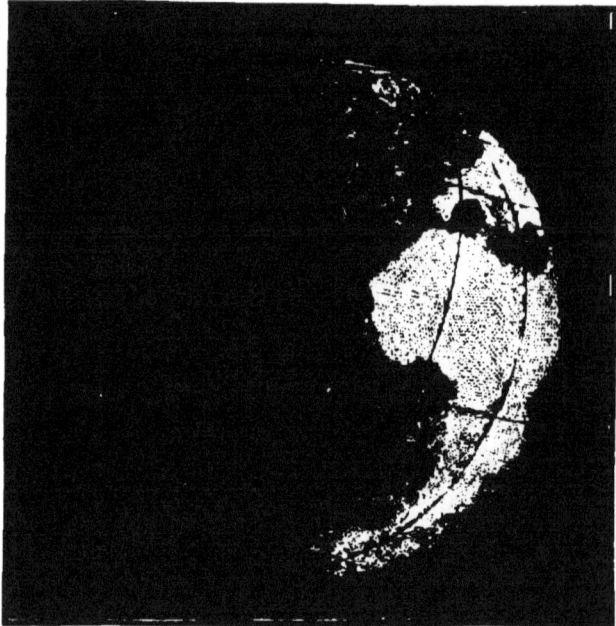

Fig. 11.—The earth at the summer solstice.

antecedent conditions, which form a continuous chain far away into the remote past; the volcanoes of olden times differed not materially from those which are now active; and the trappean rocks of Derbyshire are strictly analogous to the lava flows of Vesuvius or Etna." The phenomena of nature, as at present witnessed, thus furnish a clue to the operations of the great physical agents of the past; and this not only in geological investigation, but throughout the wide realm

of nature, and the knowledge gained will afford a stand-
point for extended survey. The great characteristic
features of the earth are thus made known, and the
forces of nature, which had mastery over our forefathers,
are becoming subservient to man. Electricity is his
obedient slave; steam arms him with giant power; with
the telescope he sounds the depths of space; and with
other instruments of precision he closely interrogates
the varied phenomena of the natural world. But his
restless spirit yearns for fuller knowledge, for wider
conquests, for more light. His aspirations the poet
has voiced :—

> " Forward, forward let us range,
> Let the great world spin for ever,
> Down the ringing groves of change."

Fig. 12.—The *Santa Maria*, in which Columbus sailed across the Atlantic.

MOUNTAINS, VALLEYS, AND GREAT PLAINS

> " Mountains that like giants stand
> To sentinel enchanted land.
> High on the south, huge Benvenue
> Down on the lake in masses threw
> Crags, knolls, and mounds confus'dly hurled,
> The fragments of an earlier world :
> A wildering forest feathered o'er
> His ruined sides, and summit hoar,
> While on the north through middle air
> Ben A'an heaved high his forehead bare."
> *Scott.*

THE face of the earth, like the face of man, has its distinct features. The great land masses and the wide oceans appeal strongly to his imagination ; but the lesser features—the seas with their islands, the lakes, and the rivers—are better within his mental grasp, and by careful study he may realise something of their importance in the scheme of nature in relation to man.

Of the lineaments of the land areas (which have as many millions of square miles as there are weeks in the year) the mountains may early claim investigation. Their majestic appearance, the value of the functions they perform, and their associations alike mark them out for special consideration by the student of nature. In the infancy of the world, when man, living an out-

door life, was more· possessed with the child-spirit than
now, after securing an existence from the fruits of the
earth, and some mastery over the savage creatures
that disputed his supremacy, he would be impressed
by the larger objects in nature around him.   If *First
dwelling amongst the mountains, their massiveness, notions of
the solitude of their summits—often untrodden by mountains.*
human foot and in consequence invested with mystery
—and their manifest influence in forming clouds, rain,
mist, and in giving birth to rivers, would arrest attention
and compel examination.

In the earliest times they were regarded as a *Mountains
refuge, as when destruction overtook the cities of the places of
plain Lot was enjoined to escape to the mountains. refuge.*
Later, their protective influence was perceived ; the
Psalmist, in his trouble, exclaims: "I will lift mine eyes
to the hills from whence cometh mine help." Biblical
references to the "holy mount" at Jerusalem are
numerous, the heights surrounding the sacred city being *Mountains
regarded as symbols of the divine protection of the Lord's and their
chosen people. The delivery of the law amidst the associa-
thunders of Sinai introduced an element of awe which tions.*
was intensified by the after scenes upon Mount Calvary.
But gentler feelings were stirred when from the top of
Mount Beatitudes was addressed a higher morality than
the world had previously known, and Tabor became
associated with the transfiguration of our Lord. According to Mr. Ruskin, "as the flaming and trembling
mountains of the earth seem to be the monuments of
the manifestation of His terror on Sinai, these pure and
white hills (of Hermon)[1] are the appointed memorials
of the light of His mercy that fell snow-like on the
Mount of Transfiguration."

The modern view favours the placing of mountains
amongst the earth's physical benefactors ; but before
investigating their uses in the economy of nature, it

[1] Mr. Ruskin regards Hermon and not Tabor as the scene of the
Transfiguration.

may be convenient to see how the great highlands are distributed, and what is the character of their most important divisions. The author of the *Gallery of Nature* writes: "Could a spectator command a view of the globe, supposing him to stand in New Holland facing the north, **General view of the highlands of the earth.** he would see on his right hand a continuous system of high mountains extending along the entire coast of America, linked with Asia by the Aleutian Isles. He would also see a chain on his left hand running along the coast of Africa, passing through Arabia into Persia, mingling there with the range that traverses Europe from the Atlantic, and merging in the mountains of Central Asia, which are continued north-easterly to Behring's Straits, and form the spine of the Old World." Thus, though these chains of mountains, when viewed in detail, appear isolated and utterly unsystematic, yet, when contemplated upon a grand scale, they seem to constitute but one immense range in the form of an irregular curve, with outshoots from it, bounding the bed of the Pacific on the north-east and west. The mountains and the plateaus which buttress them thus traverse the continents in the direction of their greatest extension; in the New World stretching meridionally almost from pole to pole, whilst the backbone of the Old World trends in the direction of the parallels of latitude. The opposite slopes of these ranges are of unequal length and importance. In America, the sharp dip is towards the Pacific; the eastern side presenting a gentle declivity, the land open and accessible for commercial intercourse with the Atlantic sea-board, and with the countries beyond.[1] The greatest mountain ranges are those rising two miles or more above the sea, as the Himalayas, the Andes, and the Alps. Of the second grade are the ranges with summits over one mile high, as

---

[1] In the Old World the trend of the mountains is east-north-east to west-south-west through a distance of 9000 miles. In the New World the direction is north-north-west to south-south-east over 10,000 miles, with, in each case, important subsidiary ranges.

the Carpathians, Apennines, Atlas, and the Pyrenees ;
whilst in the third rank are the mountains of Great
Britain—Snowdon, Helvellyn, Ben Nevis, and the Peak
Mountains in Derbyshire, which are half a mile or more
above the sea-line but fall short of a mile.   Inferior
elevations are reckoned as hills.

Ranges of mountains have far greater breadth than
is commonly supposed—an error fostered by the very
inadequate manner in which mountain ranges have, too
often, been represented upon maps.   Until within
recent years they have been depicted as of inconsiderable
breadth rising murally from the low ground.   A glance
at a good physical map of a mountainous country will
show the land rising in the form of a terrace, the
culminating peaks being the highest step.   If the length
of a range is expressed in thousands of miles the breadth
will run to hundreds : a mountain range with its flank-
ing high ground dominating, and not merely dividing
the country in which it is found.   A good physical map
—say of South America—will indicate that the range
of the Andes occupies a third of the total breadth of
the continent ; and the Himalayan range is in some
parts over 2000 miles across.   So considerable, in
fact, is the breadth that in a mountainous region a
traveller is often unable to ascertain from mere in-
spection the general direction of the range. *(The great breadth of first-class mountain ranges.)*

Amongst giant mountains the Himalayas are conspicuous.   They dominate South Central Asia, forming
the southernmost of three roughly parallel ranges that
rise from the plains of Asia.   They give rise to immense
rivers, and their valleys (which are not, however, on a
commensurate scale) are filled with huge glaciers, whilst
"a vast and unnumbered multitude" of very high peaks
pierce the snow-line, rising to the height of four or five
miles into the air, and are clothed in unmelting snow.
As this range is only just beyond the tropics, the lower
valleys may be filled with flowers and plants of hot
countries, as is well seen in the much-praised valley of *(The Himalayas.)* *(The proportions of the Himalayan range.)*

D

Cashmere.[1] The snow-line is here about three miles high, so that, in some parts of the range, there are two miles vertical of snow and ice, thus justifying the

FIG. 13.—Upper Valley of St. Gothard.

appellation of the name Himalayas, which means "the abode of snow."

[1] "Whose head in wintry grandeur towers
      And whitens with eternal sleet.
      While summer in a vale of flowers
      Is sleeping rosy at his feet."

A traveller thus describes one of the Himalayan peaks :—

" A mountain rose there almost sheer up from the Sutlej to the height of 13,000 feet in gigantic walls, towers, needles of cream-coloured granite and quartz,

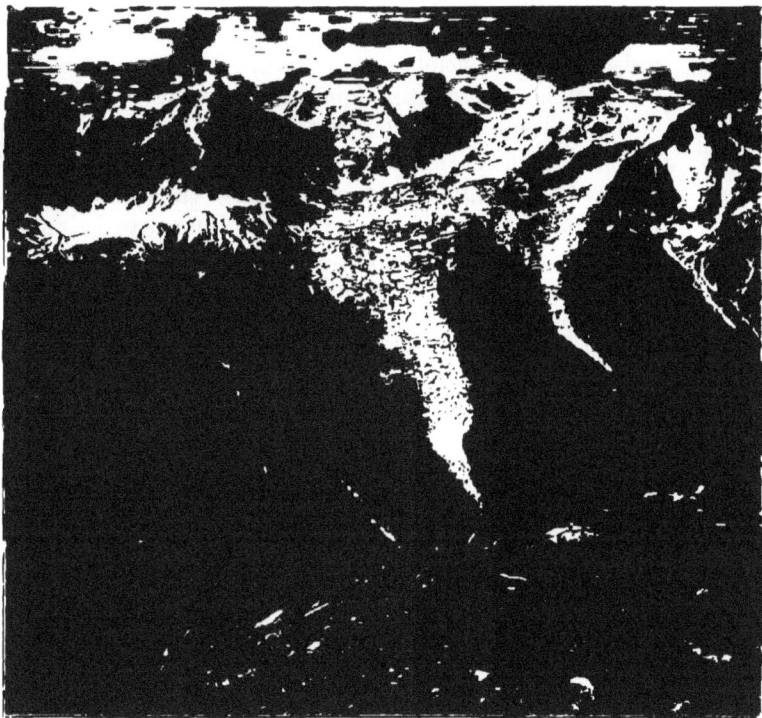

Fig. 14.—Snow-Fields and Glaciers of Mont Blanc, seen from the top of Mont Brévent.

which had the appearance of marble on which midway we stood ; whilst, itself unseen, the moon's white light illuminated the deep gorges of the river and threw a silvery splendour on the marble-like towers and battlements near.   It did not at all appear as if any external light were falling, but rather as if this great castle of the gods, being transparent as alabaster, were lighted

up from within and shone in its own radiance, throwing
its supernatural light on the savage scenes around."

The great divide of Europe. The Alps, with the Pyrenees and Balkan, form the
chief water-parting of Europe, separating the oil-using
peoples of the south from the butter-countries of the
north. Their scenery is not second to any in impress-
iveness (Fig. 13); and, situated in proximity to the
early civilised nations, they have been longer observed
and their characteristics minutely recorded.

The Alpine range has distinct names in its several
divisions, but the Pennine Alps are the most important.
Monte Rosa bounds one extremity—the eastern—and
Mont Blanc—one of the greatest of European heights—
the other.[1] In the Alpine system the great peaks rise
sharply into needle-like eminences or "horns." Their
passes or "cols" lead through magnificent scenery;
and the mountains present, at various elevations, every
kind of vegetation, from the sub-tropical flora at their
bases to the meagre plant-life of the Arctic regions
which marks the highest reaches. The snow-line[2]
ranges from 9000 to 9500 feet, the tallest of the
Alpine peaks rising far above it.

Mont Blanc. To understand the physical conditions of a mountain
like Mont Blanc is to have a key to all the rest. Mont
Blanc (Fig. 15) and Monte Rosa rise to the height of three
miles, and there are in the neighbourhood a hundred

[1] "Mont Blanc is the monarch of mountains :
They crowned him long ago,
On a throne of rocks, in a robe of clouds,
With a diadem of snow."

[2] Height of the curve of congelation :—

Place.	Latitude.	Mean Temperature of Sea-Line.	Mean Height of Snow-Line.
London . .	51½ N.	11·49 C., 52·7 F.	6,000
New York .	41 N.	16·52 C., 61·7 F.	8,138
Calcutta .	22½ N.	24·93 C., 76·9 F.	13,000

detached points,[1] each over two miles high, and all capped with snow. As the traveller descends, the scenery varies from rugged and savage around the highest points to softness and gentleness, where pine forests mingle with vineyards, gardens, and fertile feeding-grounds for flocks and herds. About a hundred years ago the first ascent of Mont Blanc was made. There are dangers and difficulties in the climb, but experience in mountaineering has been gained, and the adventurous tourist does not now hesitate to include the ascent of "the monarch of mountains" in his programme.

The monarch of mountains.

The physical exertion involved in the ascent is rendered more difficult by the air rapidly becoming rare, the atmospheric pressure being greatly reduced at the summit; and there are dangers arising from encountering the cracks or "crevasses" upon the glaciers, and from avalanches and storms of wind and snow.[2]

The summit is a crescent 70 yards long by 30 broad, covered with from 20 to 30 feet deep of snow, and the view from it is one not to be forgotten. Eastward the rival height Monte Rosa is full in view over 50 miles distant; to the west the traveller overlooks the Jura, with the plains of France beyond; to the north the Bernese Oberland, and the varied surface of Switzerland (Fig. 16); to the south Piedmont, the plains of Lombardy, and the Maritime Alps. At this lofty height the physical sensations are peculiar. The pulse rises to double its usual rate, respiration is impeded, and there is a desire to sleep. Sounds, such as the drawing of a cork or the firing of a

[1] The snow-line in Switzerland is a little over a mile and a half, and Mont Blanc is within a hundred feet of three miles in elevation.

[2] The cold encountered at great elevations is owing to: (1) the expansion of the air as it is driven up the flanks of the mountain; (2) to the want of warming from the earth, the air itself being almost transparent for heat; and (3) to the loss by rapid radiation in the dry air. The actual heat of the sun is greater on the mountain top, sometimes causing the face to be blistered.

FIG. 16.—A view in the Valais below St. Maurice.

pistol, are greatly enfeebled, and overhead the sky
loses its rich blue colour.[1] It is instructive to make the
circuit of Mont Blanc, observing the mountain from
several standpoints, and in such a tour over thirty
valleys filled with glaciers would be passed, one of the

FIG. 17.—The Matterhorn

finest being that containing
the Mer de Glace, which is
seven miles long, and which
creeps down the valley at the
rate of a foot a day. None of the Alpine peaks strikes
the traveller more than the Matterhorn (Fig. 17). Its

[1] The blue of the sky is principally due to the presence of water
vapour in the air, water even when in the state of vapour having a
blue tint when seen in mass. The traveller, as he ascends, leaves
the water vapour below him to a large extent, and the sky consequently loses something of its characteristic tint.

height is nearly three miles, the last great step of
3800 feet rising in one grand pyramid of rock with
sides so steep that the snow with difficulty finds lodg-
ment.   It impresses the traveller with the idea of
absolute permanence, yet the late Professor Tyndall,
who made the Alps a special study, says that no mountain peak gives such evidence of the destructive power
of frost and snow as the Matterhorn.

The Pyrenees also exhibit fine scenery and contain
some of the loftiest heights and grandest valleys in
Europe, as, for instance, that near Mont Perdu, which
is over half a mile deep.   Several of the passes are very
high, and so beset with danger that there is a saying
amongst the inhabitants that "the father never waits for
his son, nor the son for his father" when traversing them.

Amongst the less important mountains of the
eastern hemisphere are the "frosty Caucasus," with
noble peaks, as Elburz, which is loftier than Mont
Blanc, in addition to valleys of great beauty and
productiveness, and with no less than thirty considerable rivers.   The beech, the chestnut, and other forest
trees so abound here that one region has been called
"the sea of trees," and a traveller describes the district
as "one continued garden of azaleas, rhododendrons,
and myrtles."   The mountains of Armenia are interesting from their associations as well as from their proportions.   Here is Ararat, "the mountain of the ark,"
which is loftier than Mont Blanc by the whole height
of Snowdon.   "It is impossible," says an observer, "to
describe the effect produced by the first view of this
stupendous mountain rising in majesty and solitary
grandeur from the plateau of Armenia."

Among isolated mountains which have acquired unusual interest from their associations may be named
Mount Tabor in Northern Palestine, celebrated in the
sacred writings and admired by modern travellers.
"The heavens," says the Psalmist, "are thine, the earth
also is thine : as for the world and the fulness thereof,

(marginal note) Frosty
Caucasus

thou hast founded them. The north and the south, thou hast created them : Tabor and Hermon shall rejoice in thy name." The mountain, according to Carl Ritter, rises some 800 feet sheer from the plain of Esdraelon, its total height above the level of the Mediterranean being about half a mile. But it looks twice as high as it really is on account of its position ; its symmetrical cone rising like an altar from the plain and forming a landmark of the northern portion of the Holy Land. Tabor is clothed to its summit by luxuriant forests. Here blend the dark green foliage of the walnut, the more familiar oak, and a rich undergrowth of rose bushes, mingled with the white blossoms of the styrax. Another traveller tells of countless birds singing their morning songs, and of the thought awakened within him that "here once walked Jesus," this being the reputed Mount of Transfiguration. The view from the summit, a substantial plain a mile across, is one to be remembered. At the base spreads the green plain of Esdraelon, which has been the battle-ground of several contending armies, but which is now peaceful enough. "To the north, showing blinding white, is the snow-crowned range of Anti-Lebanon ; to the south, in the blue distance, are the Judæan hills ; and, near at hand, and clothed with a mantle of pale green, are the heights of Gilboa, memorable as the scene of Saul's last fight and of David's touching elegy. A corner of the Galilean lake gleams near, and the Moabite mountains run sharply out from the horizon like an impassable wall which, however, a nearer view would show to be rent with a thousand Titanic seams." The villages of Nain and Endor, close at hand, give human interest to this wide landscape ; westward are the wooded heights of Carmel, and if the weather should be favourable (and the constantly clear atmosphere renders an ascent of this mountain not the speculative undertaking which it would be in England), a glimpse may be obtained of the blue Mediterranean.

Margin notes:
Tabor the Mount of Transfiguration.

The view from Mount Tabor.

The eastern traveller, charmed with the brightness
and beauty of Mount Tabor, might in contrast turn to
scenes of an altogether different type—the mountains
of Sinai and Horeb. Here the aspect is rugged and
desolate, adding force to the dramatic circumstances
which accompanied the giving of the law from their *The*
summit, "when Sinai was altogether of a smoke, because *mountain*
the Lord descended upon it on fire; and the smoke *of the law-*
ascended as the smoke of a furnace, and the whole *giving.*
mountain quaked greatly, and the voice of the trumpet
sounded long and waxed louder and louder."   The
pilgrim's path is here amidst rocks of granite. There is
a tradition that the tables of the law were made of
graphic granite, in which rock there is imitation of
Hebrew characters, and that they are buried beneath
the floor of the chapel of Elijah,[1] which stands upon
the mountain. Sinai and Horeb are thus rich in
historic and legendary interest; "every little ruin,
every crumbling wall having its story."

In the western hemisphere there is a simpler scheme *The*
of highlands than prevails in the east. The great *western*
range that extends almost from pole to pole, under the *world.*
various names of the Rocky Mountains, the Sierras of
Mexico, and the Andes, forms the water-parting of this
western continent, dividing the land meridionally and
very unequally. There is a short, sharp slope through
which rapid streams descend to the Pacific; and a
gentler eastern declivity down which mighty rivers
move slowly towards the Atlantic, affording means of
internal communication of the utmost value in the
development of the natural productions of the land.

The Rocky or Stony Mountains are upon a gigantic *The*
scale. Many of the peaks are over two miles high, and *cañons*
the valleys they form on the Pacific side are stupendous *of the*
Colorado.

---

[1] Elijah the Tishbite retreated to these mountains when disheartened with the wickedness of the people in the plains around.   A
modern entry in Elijah's chapel gives the barometric pressure 21 in. ;
thermometer 13° Reaumur at mid-day.

FIG. 16.—Grand Cañon of the Colorado.

(Fig. 18). The narrow and deep gorges, or cañons, as they are locally designated, are in some instances almost inaccessible; but such travellers as have penetrated into them have been struck with the remarkable scenery they present. Through the Bloody Cañon in the **The Bloody** upper part of the Yosemite valley a river runs at an **Cañon.** enormous depth; its course is too difficult to follow, so that the traveller has to be content with imperfectly viewing it from the tableland through which the waters have cut their way. The vast elevated and terraced plains associated with the western slope of the Rocky Mountains, which are bare of trees, are half scorched by day and half frozen at night.[1] There are no wild animals, no water (except as stated, at inaccessible depths, say half a mile or a mile below the surface of the plateau); there are no sloping valleys, but crack-like depths of the most remarkable character. The river passes for 300 miles through defiles varying from half a mile to a mile in depth, forming the grandest geological section in the world. " Brilliant tints of purple, green, brown, red, and white," says a visitor, " illuminated the stupendous surfaces and relieved the otherwise sombre monotony. In places the river narrows at one point to fifty yards, where it is studded with rocks, which increase the interest of the scene; and, far above, clear and distinct upon the narrow strip of sky, turrets, spires, jagged statue-like peaks and grotesque pinnacles overlook the deep abyss. No description, however, **A mighty** can convey an idea of the varied and majestic grandeur **defile.** of this peerless waterway." " Wherever," says another writer, " the river makes a turn, the entire panorama changes, and one startling novelty after another appears and disappears with bewildering rapidity. Stately façades, august cathedrals, amphitheatres, rotundas, castellated walls, and time-stained ruins, surmounted by every form of tower, minaret, dome, and spire, have

---

[1] It should be borne in mind that in sub-tropical regions there are no short nights.

FIG. 19.—The Grand Cañon, Colorado.

been moulded from the cyclopean masses of rock that form the mighty defile. The solitude, the stillness, the subdued light, and the vastness of every surrounding object produce an impression that ultimately becomes almost painful" (Fig. 19).

The valley has not ceased to be attractive to the geologist as well as to the traveller. The Yosemite valley especially is so deep that an American scientist has suggested that it can hardly be wholly due to the erosive action of water, and that in one of the volcanic convulsions which the district has experienced the bottom of the valley has subsided. Wallace, in the *Nineteenth Century* for March 1893, furnishes additional particulars of the valley, but combats this view of its formation. The Yosemite valley is seven miles long, with a width varying from half a mile to a mile, and is chiefly formed in granite, with an admixture of metamorphic slates, tertiary gravels, and beds of lava, and has almost vertical sides.

A stupendous ravine.

"Some of these precipices are absolutely perpendicular from base to summit; and the upper part of the valley has numerous waterfalls. El Capitan, at the lower end of the valley, is a ·smooth vertical wall of granite 3300 feet high, with no visible crack or ledge upon it from top to bottom." Cathedral Rock opposite is 2600 feet; Sentinel Rock, 3000 feet; Half Dome, at the upper end, is no less than 4737 feet, the upper 1500 feet vertical; North Dome, 3568 feet, has its summit beautifully rounded, but broken below so as to show the concentric layers of which it is formed. Wallace, after duly considering the volcanic action in the valley, nevertheless holds that it is mainly the work of flowing water.

The comparatively unvisited Blue Mountains of New South Wales also exhibit some remarkable valleys (Fig. 20). A great sandstone plateau is at one point broken by deep ravines with almost perpendicular cliffs from 1000 to 2000 feet in height. Darwin, in his

The Blue Mountains.

Volcanic Islands, says concerning them : " It is not easy to conceive a more magnificent spectacle than is presented to a person walking on the summit plains. He can strike with a stone (as I have tried) the trees growing at the depth of 1500 feet below him." The calculation has been made that from one of these valleys 134 cubic miles of material have been excavated by water.

FIG. 20.—A View on the Zig-Zag of the Blue Mountain Railway.

The Sierras. The Sierras of Mexico and Central America are also of the most interesting character. The American historian Prescott incidentally brings us face to face with the physical features of this part of the world, which exhibits the variety of climate, productions, and general aspect due to alteration of height in a remarkable manner.

On the plains near the sea (the *tierra caliente*) " fruits and flowers chase one another in an unbroken circle through the year ; the gales are loaded with perfumes

till the sense aches at their sweetness." The Spanish Mexican mountains.
cavaliers on their march to Mexico passed upwards to the
tableland of Xalapa, "where the wealthy resident of the
lower regions retires for safety in the heat of summer,
and the traveller hails its groves of oak with delight as
announcing that he is above the deadly influence of the
vomito" (the malarial fever of the district). At a still
greater elevation the cold winds from the mountains
mingled with rain and driving sleet and hail, and the
country became wild and dreary. But from these
elevations they could see the fertile country they had
left below. "The path often led them along the border
of precipices, down whose sheer depths of 2000 or 3000
feet the shrinking eye might behold another climate,
and see all the glowing vegetation of the tropics."[1]

The Andean range of South America stretches for The Andes.
4000 miles close to the western sea-board, so that only
mountain streams descend to the Pacific; but down the
eastern slope rivers of gigantic proportion form and
flow. The snow-line is here 17,000 feet, or over three
miles high, and the Cordilleras (as the chains of moun- The back-bone of South America.
tains are locally called) are surmounted with peaks, the
loftiest of which are four miles above the sea, and far
above the line of congelation. The material comprising
the range is generally porphyry, or one of the earlier-
formed igneous rocks; and, as is often the case, they
abound in metallic ores, amongst which silver is chief,
the mines of Potosi yielding it having a world-wide
reputation. One of the best-known of the Andean
heights is Cotopaxi, which is described by Humboldt Cotopaxi.
as the most beautiful and regular of all the colossal
summits. It is perfectly conical, and its peak, covered
with snow, shines with dazzling splendour. In elevation Cotopaxi equals Vesuvius piled upon Etna, and Snowdon upon Vesuvius, being $3\frac{1}{2}$ miles above the sea-level.

The uses of mountains in the economy of nature are
numerous and important.

[1] *History of the Conquest of Mexico.*

E

Uses of
mountains.

"They direct," says Mr. Ruskin, "the form and flow of rills, streams, and rivers; they assist in changing the currents of air, and so give circulation to the atmosphere, man's vital fluid; they give change to the ground, breaking up the otherwise monotony of the earth; and they change and enrich the soil. Mountains may be regarded as heaps of fertile and fresh earth, laid up by a prudent gardener beside his garden beds, whence at intervals he casts on them some scattering of new virgin ground." By every rain shower the substance of the mountains is washed down to enrich the low-lying lands, and "the winter floods, which inflict a temporary

Mountains
beneficial.

devastation, bear with them the elements of fertility; the fruitful field is covered with sand and shingle in momentary judgment, but in enduring mercy, and the great river which chokes its mouth with marsh, and tosses terror along its shore, is but scattering the seeds of the harvest of futurity, and preparing the seats of unborn generations."

The etymology of mountain (Greek *meno*, to remain) is suggestive of their permanence. They form natural boundaries between continent and continent, country and country, and by clearly defining territories prevent disputes and conduce to civil order. No other line of division is needful where there is a mountain barrier. Mountains are also convenient highways into the higher strata of the atmosphere, where some important problems in science may be solved. It was at the top of the Pûy de Dome that Pascal made his crucial test upon atmospheric pressure, and showed that the barometer falls at a constant rate according to altitude;[1] and other

---

[1] Ben Nevis has a meteorological observatory (Fig. 21) 4407 feet above the sea, whose records are published daily in the press. In the *Times* of 28th March 1893 the barometric reading is given at 25·623 inches at 9 A.M., and 25·594 at 9 P.M., whilst the record at the base station for those hours was 30·248 and 30·230. Particulars are also given of the temperature and the wind, sky, and hygrometric state. Upon the loftier Mont Blanc a weather observatory is also established.

FIG. 21.—Ben Nevis Observatory.

questions in physics (as in light, heat, the play of the electric forces, and the power of gravitation) may, from the position of advantage of a mountain summit, be asked and answered.[1]   A mountain range, even if barren, may convert an otherwise sterile region into a scene of fertility ; for although mountainous districts are not in themselves the most favourable for either dairy farming or tillage, they are the means of suc-

Mountains a cause of fertility. cess in both.   The water vapours precipitated in rills and streams carry amenity to the lower ground, assist drainage and irrigation, and promote internal communication, enabling a country to exchange its products with advantage.   The glaciers of the higher valleys help to feed the rivers and preserve a constant flow.   Unusual heat in the plains tends to dry up the rivers, but melts the glaciers in a proportionate degree, a method of compensation well seen in the Ganges and the Rhone, which are glacier-fed rivers.   Mountain regions are proverbially rich in minerals.   The earth's metallic treasures are thrown up near the surface by the convulsions which have raised the mountains them- selves, or the weathering of neighbouring exposed points ; and "the precious things of the lasting hills" are rendered accessible to man.   The mining countries of Europe are the mountainous ones ; and, amongst the hills of Cornwall, Wales, and Derbyshire the mining industry of England is profitable.

Then mountain countries are the natural homes of freedom.   There seems to be something in the very air of mountain regions which fosters liberty.   The in- habitants have a proud consciousness of independence, the natural birthright of the child of the mountains. The mountaineer has enormous advantages for defence against his enemies.   Passes are "gates of the hills," and a narrow defile is easily defended by few against

[1] The Spectre of the Brocken in the Hartz Mountains, and the appearance of the rainbow, forming an almost complete circle, may be instanced.

many.   In Macaulay's *Lays of Ancient Rome* a noble
soldier is made to say—

> "I with two more to help me
> Will hold the foe in play :   .
> In yon strait path a thousand
> May well be stopped by three."

The Rock of Gibraltar is a natural fortress (Fig. 22)   Mountains
guarding the entrance to the Mediterranean.   as fortresses.

Fig. 22.—Gibraltar.

English history also offers illustrations.   The Roman
invaders found great difficulty in conquering Scotland,
and they never subdued Wales.   The Welsh also withstood the Saxons, and for a long time the armies of
consolidated England.   The Swiss in their mountain
home have maintained their freedom inviolate despite
the powerful nations by which they are surrounded;
and Schamyl, the Circassian chief, from his garrisons
amongst the Caucasus, beat back again and again the
vast armies of Russia.   The Afghans also, fighting
amongst their native hills, have been found difficult to
subdue, and at the Kyber Pass inflicted a disastrous

defeat upon the British arms.   Similarly, the Aztec
warriors of Central America made a strong defence [1]
against the disciplined chivalry of their European in-

vaders; and the Sierras of Spain also played an important part in the struggle against the Moors.   Thus
"when the Spaniards quitted the shelter of their
mountains and descended into the open plains of Leon
and Castile, they found themselves exposed to the pre-
datory incursions of the Arab cavalry, who, sweeping
over the face of the country, carried off in a single foray
the hard-earned produce of a summer's toil"; and "it
was not until they reached some natural boundary, as
the river Douro, or the chain of the Guadarrama, that
they were enabled, by constructing a line of fortifications
along their primitive bulwarks, to secure their conquests,
and oppose an effectual resistance to the destructive
inroads of their enemies."   That mountains greatly
enhance the beauty of the earth is a trite observation.
A background of mountains is an essential feature of a
perfect landscape ; and the poet and the painter derive
inspiration from their presence.

"Then the mountains how fair !
Towering up in the sunshine and drinking the light,
Whilst adown their deep chasms all splintered and riven,
Fall the far-gleaming cataracts silvery white."

A competent authority has suggested that mountains
are not without their influence in improving the type
of human life ; their magnitude rendering man more
receptive of great ideas, their steadfastness inciting him
to constancy, and their beauty having a refining effect.

To write an epic poem, said the ancient philosophers, a
man must have lived amongst the mountains.   Ruskin
has thus told of the effect of his native mountains upon
the French philosopher Pascal :—

"Born at Clermont in Auvergne, under the shadow

[1] As at Bishop's Pass, Mexico, "capable of easy defence against an
army."

of Pûy de Dome, though taken to Paris at eight years
old, he retains for ever the impress of his birthplace.
Pursuing natural philosophy with the same zeal as
Bacon, he returns to his own mountains to put himself
under their tutelage, and by their help discover the
great relations of the earth and the mountains. Struck
at last with mortal disease, gloomy, enthusiastic, and
superstitious, with a conscience burning like lava, and
inflexible like iron, the clouds gather about the majesty
of him, fold by fold; and with his spirit buried in
ashes, and rent by earthquakes, yet fruitful of true
thought and faithful affection, he stands like that mound
of desolate scoria that crowns the hill ranges of his
native land, with its sable summit far in heaven, and
its foundations green with the ordered garden and the
trellised vine."

The mountain ranges of the torrid zone by their
elevation increase the habitability of the earth.   In
India there is, at the Sanatorium of Darjeeling, high up
on the slopes of the Himalayas, an escape from the heat
of the northern plains; and the dweller in mountainous
countries of the tropics may obtain almost any climate
he may desire at will.   In the continent of South
America this effect may be well studied, the tropical
plains experiencing constant hot weather, whilst spring
and winter are each permanently enthroned at a definite
height above the sea.   In Switzerland also the tourist
may, in a few hours, exchange the vineyards of the
lower valleys for the frost-laden breezes of the mountains.   In England the mountains are not high enough,
and the latitude not low enough for these changes to
be very marked; but, even in this country, the bearing
of elevation may be noticed.   Snow is rarely absent
from the higher valleys of Snowdon; and of the loftiest
of the hills of the Peak in Derbyshire there is a
countryside saying—

> " If there be snow without
> It will lie on Kinderscout."

Mountain districts as sana-toriums.

Buxton, in the same district, is the highest considerable town in the British Isles. The lowest parts of the valley in which it stands are a thousand feet above the sea, and the atmospheric pressure is reduced an inch with important effects upon health ; and, by ascending the neighbouring mountains, an elevation of nearly half a mile may be reached.[1]

Amongst the lesser uses of mountains may be mentioned their employment, anterior to the advent of our telegraphic system, as beacons and signal stations, a mode of communication which could be again easily revived in the event of an enemy cutting the telegraphic wires, or their being injured by storms. When England was threatened by the Spanish Armada the beacon fires from their summits flashed the alarm :—

Mountains as beacons.

"Southward from Surrey's pleasant hills flew those bright couriers forth ;
High on bleak Hampstead's swarthy moor they started for the north ;
And on and on without a pause, untired they bounded still ;
All night from tower to tower they sprang, they sprang from hill to hill ;
Till the proud Peak unfurled the flag o'er Derwent's rocky dales,
Till like volcanoes flared to heaven the stormy hills of Wales ;
Till twelve fair counties saw the blaze, on Malvern's lonely height,
Till streamed in crimson on the wind, the Wrekin's crest of light."

It may be remembered that many of the beacon fires were lighted as part of the commemoration proceedings at Her Majesty's Jubilee.

[1] If the atmospheric pressure upon the body of a man of average size be taken at 14 tons at the sea-level (reckoning 14½ lbs. per square inch of surface exposed), then at Buxton the pressure upon him would be reduced by half a ton.

Fig 23.—Gorge of the Ericht, above Blairgowrie.

III

SCENERY AND ITS CAUSES

" The sounding cataract
Haunted me like a passion ; the tall rock,
The mountain, and the deep gloomy wood :
Their colours, and their forms have been to me
An appetite."

Early notions. How soon primitive man began to be impressed by the appearance of objects around him in nature is not easy to determine. In the "childhood of the world" he would be largely occupied in providing for mere existence ; in defending himself against the beasts of the field, many of which surpassed him in physical endowment, and were not much his inferior in intelligence ; and in gathering from the wild fruits a meagre subsistence. As soon as some mastery of the position had been attained there would be opportunity for observing more closely the prominent natural features of the vicinity, as a mountain range, a large lake, or river, which would arrest attention and invite examination. Perceiving that they embodied a power greater than their own, the early dwellers upon this earth were led Natural objects personified. to personify and worship natural objects. Thus the river Danube was regarded as an emblem of magnitude and energy ; the Ganges was esteemed sacred, a superstition that has not nearly passed away ; and every hill, every forest, every lake was looked upon as the workmanship and shrine of a divinity. A

patriotic Roman soldier is made by Macaulay to
exclaim—

> "O Tiber ! father Tiber !
> To whom the Romans pray,
> A Roman's life, a Roman's arms
> Take thou in charge this day."

The Ethiopian held that God made *his* sands and
deserts while angels were employed in forming the rest
of the globe. The pen of Virgil is fluent in description
of landscape beauty ; in the odes of Horace this theme
is expanded ; and Tasso, who lived in the fertile plain
of Lombardy, which is watered by the noblest of Italian
streams, delighted to portray the scenery of his native
land. *Natural objects venerated.*

As mankind became more observant of the features
of the landscape, new ideas would seek expression, and
new words and phrases would be added to the vocabu-
lary. It is noteworthy that words in the final form
relating to scenery are derived from several sources.
Thus, from the Welsh comes garden ; from the Gothic,
dale ; and from the Danish, lawn. Our Saxon forefathers were keenly appreciative of landscape beauty,
some of our most expressive words being of native
origin, such as dingle, hill, field, meadow, stream, flood,
and sea. To the Normans the language is indebted for
wood, river, cascade, vale, rock, and forest ; and to the
Romans for rill, valley, ocean, and lake. *Words expressing scenery.*

The ancient schools of philosophy were not without
teachers who perceived that the scenery of the country
was not permanent—that the mountains had a beginning, a growth, and would have an end ; that a river in
its flow was not perpetual—that, in fine, the present
scene of things is only a phase in the earth's history.
The earth a scene of change.
But the ideas of the majority would well accord with
those expressed in an Arabian allegory : "I passed one
day by a very ancient and populous city, and asked one
of its inhabitants how long it had been founded. 'It is
The truth in an allegory.

indeed a mighty city,' replied he; 'we know not how long it has existed, and our ancestors were, in this respect, as ignorant as ourselves.' Some centuries afterwards, as I passed by the same place, I could not perceive the slightest vestige of the city. I demanded of a peasant who was gathering herbs upon its former site, how long it had been destroyed. 'In sooth, a strange question,' replied he; 'the ground has never been different from what you now behold it.' 'Was there not,' said I, 'of old, a splendid city here?' 'Never,' answered he, 'so far as we know, and never did our fathers speak to us of such.'" Other changes occur, with new witnesses who are ignorant of previous conditions.

But with advancing knowledge, says the geologist, man's view of nature took a different course. "The gods were dethroned, and the invisible spirits of nature no longer found worshippers; but it was impossible that the natural features which had prompted the primeval belief should cease to exercise a potent influence on the minds of man. This influence has varied in character from generation to generation, as is seen by comparing the literature of successive periods. Probably at no time has it been more potent than it is at the present day."

The face of nature is, indeed, a transformation scene; but the changes are so slow that the life of man is too short to appreciate the mutability; yet operations that seem insignificant in themselves by constant repetition carried on for centuries become mighty agents of change; and it is clear enough, from the testimony of the rocks, that *the scenery of the earth is the net result of the working of the opposing forces of waste and renewal.*

English landscape.   In a small country like England, and which has a middle position in latitude and in height above the sea, natural phenomena are not so impressive as in the tropics, or in Polar or elevated regions. Its geological structure, upon which scenery depends, is, however, so varied that we have, upon a small scale, almost every

FIG. 24.—Group of beeches, Burnham.

kind of scenery. " England," says Emerson, in his *English Traits*, " has a singular perfection. In variety of surface it is a miniature Europe, having plain, forest, marsh, river, sea-shore ; mines in Cornwall, caves in Derbyshire,

Variety of scenery in Britain. delicious landscapes in Dovedale, and sea-views at Torbay ; highlands in Scotland, Snowdon in Wales ; in Westmoreland and Cumberland a pocket Switzerland, in which are lakes and mountains on a sufficient scale to fill the eye and touch the imagination." Scenery, as stated, depends upon geological character, and no country, for its size, includes so many different formations as the British Isles. The great forests which formerly covered the land have been largely removed, being now forests only in name, and the oaks of Sherwood Forest and the beeches of Burnham (Fig. 24) are to be regarded as the relics of an age that is past.

Interesting geological structure of England. The land has been formed piecemeal, as it were, by the orderly working of nature's forces. Very early the slate rocks of Wales, which were formed of the deep mud of the coast of primeval lands, were laid down ; then followed the coal-fields and sandstone tracts of the Midlands, spread out along the shores of hardly less ancient waters ; and, subsequently, the chalk downs of the south-east were formed from the oozy bed of a deep sea. These rocks (which are by no means a complete list) bear record of their previous state. The chalk hills, for example, are full of the fossil remains of marine creatures. The sandstones bear ripple marks of the waves ; and strange beasts that lived upon the coasts left their footmarks on the wet sands. The changes that have been wrought are partly due to destroying agents which have worn away the land, or to the restoring work of other forces, and thus the scenery of the world has been fashioned.

Agents of waste. The action of the sea is a principal cause. The shaping of the continents, the outline of Europe, and the long irregular coast-line of England are principally due to the work of the waves, the currents, and the tidal

FIG. 25.—The Armed Knight and the Long Ship's Lighthouse, Cornish coast;
illustrating bombardment by the sea.

Marine bombardment.

The coastline carved by the sea.

movement of the ocean. The openings and the promontories afford a study of the attack of the sea and the defence offered by the land. The bays are where the softer strata have yielded to the wash of the sea; the headlands where harder strata have withstood Neptune's assault (Fig. 25). The coast of Wales and of the west of Ireland and Scotland well illustrate this action, being exposed to heavy storms from the Atlantic. Amongst the Hebrides, after a tempest, fragments of rock are scattered about like chippings in a mason's yard; and these rock fragments become ammunition for the artillery of the sea. On the east coast of England the attack is less severe; but the rocks are of less enduring material, and marked effects are produced. On the Yorkshire coast several villages have been devoured by the sea—Ravenspur, for example, where Bolingbroke landed to claim the throne of England, is not now to be found; its name only remains upon ancient maps. Outlying rocks and islands near the coast everywhere indicate how, in time past, the sea fought its battle with the land; and these outposts will, unless the restoring agencies of nature should come to their aid, be ultimately captured and overthrown; for "the waters wear the stones," and the most enduring of them must give way to the constant sapping of the sea. The line of caverns often found upon a rocky coast, between high and low water, also tells of the play of the sea; the rocks are sculptured into long flowing curves, which are the characteristic tool-marks of the waves. The seas around our shores are no less tempestuous than when they were encountered by the hardy Norsemen in their piratical descents upon England; the tidal action is as marked now as when it excited the wonder of the Roman invaders (the Italian coasts hardly showing any movement of the tides), and there is no evidence that the sweep and power of the currents has diminished in any material degree. The British Isles have two thousand miles of coast-line, upon which, as in all maritime countries, tho sea is acting night and day with unceasing perseverance and with specially destructive effect during storms.

Another cause of changing scenery is frost. In England its power is feeble. But in Arctic lands and on lofty mountains in all parts of the world its action is powerful and constant. In the high Alps thero is a frost every night in tho year. Water imprisoned in

The destructive power of frost.

Fio. 26.—The great Báltoro Glacier, Himalayau Mountains.

crevices of the rocks expands with immense force,[1] and breaks up the face of the mountain. Even in England, after a spell of frost, we may occasionally see at the foot of a cliff chips of rock that the frost has broken off. It is, however, amongst the mountains that pierce the snow-lino that the maximum result is perceived. A traveller in the Himalayas writes: "The whole mountain-side was covered for a long way with huge blocks of granite and gneiss, over which we had to scramble,

The force of frost in England.

---

[1] If a small cast-iron bottle, fitted with an iron screw-stopper, be filled with water and placed in a freezing mixture, the water will freeze and burst the bottle.

F

A stone
avalanche.

and here I made my first experience of what a granite
avalanche means, but should require the pen of a Bunyan
to do justice to the discouraging effect on the pilgrim."
The broken faces of rock upon which the play of light
and shade is so pleasing are thus largely due to the
wasting effects of frost.   The mountains are gradually
demolished, and the glaciers carry down the fragments
of rock which fringe their sides (Fig. 26) to form mounds
or moraines in the lower valley, where the glacier dissolves and dies.   Such masses of débris are found in
this country, for its northern half once lay, as Green-

Ice work.

land lies now, under a thick ice sheet.   In constructing
the branch railway from Settle to Carlisle several of
these moraines were cut through.   Frost is also
accountable for the erratic blocks scattered over parts of
England, north of the Thames.   These are foreign
stones brought by glaciers from the hill-country around,
or by icebergs when the districts where these blocks

Icebergs
the carriers
of nature.

occur lay under the sea.   Icebergs are huge fragments
of glaciers from the valleys where they are formed,
broken off upon entering the sea, and bearing fragments
of the rocks, often of considerable size, which have fallen
from the mountain-sides.   As these floating ice mountains dissolve in the warm seas into which they are
borne by the ocean currents, their cargo of rocks is
discharged upon the sea-floor; and when—it may be
long ages afterwards—the ocean bed is raised to the
dignity of dry land by the upheaving action of submarine
fires, these travelled blocks are brought to the light of
day.

Weather-
ing action.

The action of weathering also modifies scenery.
Rain, especially when driven by the wind, takes a
prominent part in this work.   The rocks are gradually
destroyed (Fig. 27), and the softer materials of the earth's
coast are carried, as turbid water, towards the lower
ground, and the hardest rocks have in time to yield.
Upon limestone rain-water has a solvent as well as
mechanical action, so that in tracts of this formation, as

Fig. 27.—Brig o' Trams, Wick. Cliffs of Old Red Flagstone, showing how the inward inclination of the precipices is determined by lines of joint, and how the faces of rock are etched out by weathering along the lines of bedding.

The weather a reducing agent.

in North Derbyshire, valleys are extensively excavated, and caverns (many of which are valleys in the first stage) are numerous and characteristic. In the tropics this action acquires its greatest intensity, the rain falling with a violence unknown within the temperate zone. At Calcutta, for example, as much rain has

Scenery caused by rain.

been registered in one day as falls in England in a year; and amongst the hill ranges which skirt the Malabar coast on the west of India[1] the rainfall is the heaviest known. The Sivalik Hills, in the Himalayan region, present fine illustrations of rain action. Large blocks of stone abound in the valleys, and these have protected the ground they cover, whilst, all around, the earth is worn away by the rains, the stone at length forming the capping of a mound.

The rains feed the rivers, which are prime factors in slowly changing the aspect of a district. They steadily carry the mountains into the plains, and bear away the spoil of the continents into the sea.[2] English rivers, although more important than their size would indicate, are the merest streams compared with the giants of the great land masses; and the untravelled Englishman with difficulty grasps their magnitude and influence. The

Rivers as agents of change.

Amazon has 50,000 miles of navigable waters in its system, and its estuary is like an arm of the sea. Upon the Mississippi the traveller may steam for 2000 miles without perceiving much diminution of width; and the mouth of the La Plata is 100 miles across, the influence of its stream being felt some hundreds of miles away at sea. But the finest example of transporting power is furnished by the Ganges. It is computed that, during

[1] On the opposite side is the Coromandel coast, which may be remembered by its being on the same side as Calcutta, both names having the initial C.

[2] If the average elevation of the American continent be taken at 748 feet, as given by Humboldt; and if the rate of destruction by the rivers be a foot in six thousand years, it is clear that, in four million years, this continent would be entirely carried away if no elevatory forces came into action. The wearing would, however, steadily decrease as the slopes became less.

the flood season, this river bears to the sea, every day, material sufficient to form a hundred of the largest pyramids. Again, the land of Egypt is "the gift of the Nile," this fertile valley, 1000 miles in length, having been constructed by the river of soil brought from the highlands of Abyssinia.

New low-lying land is also forming at the mouths of many rivers at a rapid rate. The Zambesi, flowing through equatorial Africa to the Indian Ocean, has spread out a delta the size of Scotland ; and, nearer home, the rivers of England are slowly converting their estuaries into deltas by the material they transport and deposit, and which action would be more rapid and regular but for the tidal scour.

On the other hand the igneous forces of nature tend to restore the balance. Volcanoes eject streams of lava miles in length and mountains of ashes which rebuild the wasted land. The earthquake thrust, and the steady rise of the land observed in certain districts have the same effect of withstanding the wearing energies of the earth and preserving equilibrium. The wind and the sea have also, it should be remembered, a restoring action, as is well seen on the Lancashire coast between Liverpool and Fleetwood, where sand borne inwards by the wind and the waves has formed new lands, Southport standing upon sand so formed. These operations produce distinct types of scenery as the constructive or the destructive forces prevail.

Of the interest and beauty of the scenery amongst the Himalayan Mountains, Mrs. Somerville has told in her fine treatise on *Physical Geography.* The loftiest peaks being precipitous, and therefore bare of snow, give sublimity to the scenery of these passes. ^{Types of scenery.}

"During the day the stupendous masses of the mountains, their interminable extent, their variety, and, above all, the clearness of their distant outlines melting into the pale blue sky, contrasted with the deep azure above, is described as a scene of wild and ^{Mountain scenery.}

wonderful beauty.   At midnight, when myriads of stars
sparkle in the black sky, and the pure blue of the
mountains looks deeper still below the gleam of the
earth and the snow light, the effect is of unparalleled
grandeur, and no language can describe the splendour
of the sunbeams at daybreak streaming through the
high peaks, and throwing their gigantic shadows on the
mountains below.

"There, far above the habitations of man, no living
thing exists : no sound is heard ; the very echo of the
traveller's footsteps startles him in the awful solitude
and silence that reign in these august dwellings of everlasting snow."

The scenery of great plains is equally distinctive.
In Europe an extensive tract of level ground stretches
through the northern part of the continent, with but
little interruption, from London through the low
countries, including the Germanic and the Baltic plain,
and the Tundra of Northern Russia onwards to the
Ouralian Mountains ; and beyond this range, which is
The of no great elevation, the plain is continued.   This flat
European land acquires a special character in the "Steppes" of
plain. Little Russia.   It has been remarked that, inhabiting
so level a country, a Russian has some difficulty in
adequately understanding the significance of a mountain
range, whilst a native of Western Europe, from his
experience of undulating scenery, is equally unable to
grasp the extent of a really great plain such as is found
in South-Eastern Russia.

Scenery of   The ground is but slightly higher than the Black
the Russian Sea—about 60 feet—and extends, with some alteration
Steppes. in character, onwards to the Arctic circle, an enormous
level broken slightly by inconsiderable hills and crossed
by numerous rivers.   The flatness is unrelieved by the
presence of trees or enclosures.   There are no roads,
for no roads are needed, the natives easily crossing in
any direction, a favourite means of travelling being by
means of a light cart with high wheels.   In slight

hollows are the Cossack villages, with posting houses where horses may be changed. The soil is often deep and rich, and, in some parts, large crops of wheat are produced. The progress of the seasons is, especially in the uncultivated parts, strongly marked. In spring the ground is gay with flowers and green with grass. In June the steppes become parched and brown beneath the fierce sunshine. The days are now long in the northern plains, so that there is an accumulation of heat, and the crops grow apace. But in many parts both rain and dew are wanting; and if the springs fail the cattle die of thirst. There is a strong saline efflorescence throughout the steppes, and other indications that the sea once occupied these great levels. In October winter begins in earnest, and soon the steppes are a wide expanse of trackless snow. Ordinary landmarks disappear. When the storm winds drive the deepening snows, it is perilous to be caught in the open, and there are records of loss of life from the rigour of the winter, as during the Crimean War, when a whole army corps perished in the attempt to cross.

At this season these plains are, however, a strong military defence. The great army of Napoleon was overwhelmed in their snows; and one of the Russian Czars had a saying that there were two generals upon whom he could rely should others fail him—January and February.

Scenery of a widely different type is presented by the forest plains of South America and equatorial Africa. Inhabitants of disafforested England are unable to comprehend the extent to which a tropical forest dominates the scene. The Silvas of the river Amazon and its tributaries extend over an area of 1500 miles in length by 800 miles in breadth—a region six times the size of France. The soil is a deep vegetable mould, the growth of centuries. In this primeval forest the trees are as beautiful in flower and foliage as they are multitudinous. They have not inconspicuous flowers like

Scenery, forest plains.

ordinary English forest trees — the oak, elm, and
ash—but rather resemble those which, exceptionally,
have showy flowers, as the horse-chestnut and the
mountain-ash. The climbing plants and the underwood
are as characteristic and important in the obstacle they
offer to the traveller's progress as the larger trees.
There is ordinarily a solemnity and solitude in the heart
of a great forest that must be experienced to be understood ; but, on the other hand, a storm in a forest of
the tropics is very impressive. Mrs. Somerville thus
writes of the Silva of the Amazon :—

"A deathlike stillness prevails during the heat of
mid-day ; but at sunrise, and sometimes in the night,
the songs of strange birds and the cries of invisible
animals give some degree of animation to the leafy
wilderness. The anxiety and terror of the animals
before a thunderstorm is excessive, and all nature
seems to partake in the dread. The tops of the lofty
trees rustle ominously, though not a breath of air agitates
them ; a hollow whistling in the high regions of the
atmosphere comes as a warning from the black floating
vapour ; midnight darkness envelops the ancient forest,
which soon after groans and creaks with the blast of the
hurricane."

An African tropical forest. A still more impressive representation of forest
scenery is to be found in the works of African explorers
who have forced a passage through the virgin forest.
In Central Africa the natives variously speak of the
jungle, the forest, the tangled jungle, and of other
modifications of woodland scenery for which the English
language has no exact equivalent. Mr. Stanley did not
find these forests silent. "The hum and murmur of
hundreds of busy insects can be heard making populous
the twilight shadows that reign under the primeval
growth " ; and other characteristics of the tropical forest
Tropical forest scenery. the same explorer has graphically set forth : " Imagine
the whole of France and the Iberian peninsula closely
packed with trees varying from 20 to 180 feet high,

FIG. 28.—The High Woods of Jamaica.

whose crowns of foliage interlace and prevent any view
of the sky and sun. Then from tree to tree run
cables from 2 to 15 inches in diameter up and down,
in loops and festoons and W's and badly-made M's,
folding round the trees in great tight coils, until they
have run up the entire length like endless anacondas. Let
them flower and leaf luxuriantly, and mix up above with
the foliage of the trees to hide the sun ; then from the
highest branches let fall the ends of the cables, reaching
near the ground by hundreds with fringed extremities,
for these represent the air roots of epiphytes. . . . Groups
of trees stand like columns of a cathedral, gray and
solemn in the twilight." After weeks of forest experience the scenery was still strange and impressive.
In the midst of the "stately forest kings," an awe
rushed upon the soul and filled the mind, which
was intensified when a thunderstorm broke upon the
scene with the most vivid lightning and deluges of
rain. Of this feeling of awe induced by tropical forests
Kingsley speaks in his *At Last*, when describing the
"High Woods of Jamaica" (Fig. 28). Many tropical or
sub-tropical islands, as the Bermudas (Fig. 29), are
equally rich in their vegetation.

Of "dark Hercynia's wood," of the Black Forest, of
the Thuringian Forest, and of the "lemur-haunted"
Hartz Woods there is here no room to enlarge. In
England there are still relics of the woods which overspread the land when our Saxon forefathers tended their
herds of swine in the forest, and bands of outlaws led
their merry life "under the greenwood tree." The
roads were then but forest tracks, and the towns were
An early enclosures in their midst. In the story of *Ivanhoe* such
English an early English forest is sketched :—
forest.
"Hundreds of broad-headed, short-stemmed, wide-
branched oaks, which had witnessed perhaps the stately
march of the Roman soldiery, flung their gnarled arms
over a thick carpet of the most delicious greensward.
"In some places they were intermingled with beeches,

FIG. 29.—Entrance to the "Convolvulus Cave," Walsingham, Bermudas.

hollies, and copsewood of various descriptions, so closely
as totally to intercept the level beams of the sinking
sun; in others they receded from each other, forming
those long sweeping vistas, in the intricacy of which the
eye delights to lose itself, while imagination considers
them as the paths to yet wilder scenes of sylvan
solitude."

Sherwood Forest. Such scenery may to some extent be yet enjoyed in the forest of Sherwood, where are interesting studies of the English trees. Many are very old
and in ruins, but hundreds of fine oaks, beeches, firs,
and birches may be seen in a twenty-miles' drive along
the "ribbon roads, the pride of Nottinghamshire"; and
the glades are more open, with a wider woodland
prospect, than in the days of Robin Hood, when the
dense growth would limit the view.

Desert plains. In striking contrast with the scenery of the woodland plains are the deserts of Africa, Asia, Australia,
and America (Fig. 30). The Sahara of Northern Africa
has been designated "the world's oven," but it is not
the uniform waste of sand, uncultivated and un-
cultivable, as is popularly understood. The western
portion, extending to the Atlantic, is marked by low-
lying tracts of shifting sand. Proceding inland east-
wards an extensive plain is reached at a somewhat
higher level, which is stony and arid, with occasional
ravines and springs of water. The central portion is
more broken, but its geography is not well known. In
some places there is no sand whatever, only hard-baked
earth with blocks of stone and pebbles. In other parts
enormous banks of loose sand are sometimes met with,
400 feet deep. In Fezzan, mountains cross the plains,
and the country shelves down to the Mediterranean
in gravelly terraces. The sandy downs which abound
in the Saharan landscape are rounded, smooth, sterile,
and dark; but in the moonlight they shine as if
Fertile spots. phosphorescent. The eastern and the northern portions
have oases or fertile hollows, with occasional rain,

Fig. 30.—The Great American Desert.

with streams and valuable productions. Here the date palm and other fruit trees flourish, and useful vegetables are cultivated with success. An oasis has generally the form of a narrow valley, sometimes 100 miles in length by a mile in breadth. An African traveller gives the following description of one of the smaller oases : "A little green corner, fresh and shady, cheered with the song of birds, and enlivened by the murmur of waters. The dates waved their elegant plumes high in the air; the pomegranates and fig trees crowded between the columns of the palms; wheat and barley clothed the soil with verdure ; the water flowed in every direction, and the humid vapours vivified the foliage. One could not help trembling for the little spot; it seemed such a feeble thing in the immensity of the desert, surrounded by desolate plains and menaced by moving sand-hills."

An oasis of the Libyan Desert.   A striking picture of an oasis in the Libyan Desert is given in Kingsley's *Hypatia* : "The great crimson sun rose swiftly through the dim night-mist of the desert, and as he poured his glory down the glen, the haze rose in threads and plumes, and vanished, leaving the stream to sparkle round the rocks, like the living, twinkling eye of the whole scene. Swallows flashed by hundreds out of the cliffs, and began their air-dance for the day ; the jerboa hopped stealthily homeward on his stilts from his stolen meal in the monastery garden ; the brown sand-lizards underneath the stones opened one eyelid each, and, having satisfied themselves that it was day, dragged their bloated bodies and whip-like tails out into the most burning patch of gravel which they could find, and, nestling together as a further protection against the cold, fell fast asleep again ; the buzzard, who considered himself lord of the valley, awoke with a long querulous bark, and, rising aloft in two or three vast rings, to stretch himself after his night's sleep, hung motionless, watching every lark which chirruped on the cliffs ; while from the far - off

Nile below, the awakening croak of pelicans, the clang
of geese, the whistle of the godwit and curlew, came
ringing up the windings of the glen, and, last of all,
the voices of the monks rose, chanting a morning hymn
to some wild eastern air; and a new day had begun
in Scetis, like those which went before, and those which
were to follow after, week after week, year after year,
of toil and prayer as quiet as its sleep."

Whilst the surface of the earth exhibits such a Under-
variety of appearance it must be remembered that some ground
of the great forces of nature which have fashioned scenery.
scenery are also active below the surface. In caverns
and in deep places of the earth the aspect is often
remarkable, and the causes of subterranean scenery Subter-
easier to determine. The play of igneous action, the ranean
circulation of water, and the delicate work of crystallisa- formative
tion gives rise to marked effects. The abstraction of forces.
enormous quantities of mineral matter to form lava-
streams and volcanic mountains has left great vacuities
in the earth's crust, and the mechanical and solvent
action of water flowing through the strata, together
with other causes, have formed caverns of considerable
extent, sometimes presenting scenery of remarkable
beauty. The granite ranges of Norway and Sweden
contain numerous caverns and fissures of great extent;
and the caverns of the Hartz Mountains, of Antiparos,
of Adelsberg, and others have been often described.
In England the towns of Buxton, Matlock, and
Castleton,[1] in the Peak of Derbyshire, have numerous
fine caverns, as have also the limestone regions of
Yorkshire and Devonshire.

It is not so much with those that have an interesting story to tell of prehistoric times, as the Victoria
Cave of Yorkshire, or Kent's Cave in Devonshire, that

[1] The new railway through the Peak makes this town, which
is so interesting to the student of nature, more accessible. Close to
Castleton is Bradwell, with fine stalactite caves, very little known;
and at Eyam, in the same neighbourhood, several caverns have recently
been discovered.

we are here concerned; but rather with those that exhibit scenery of a special character, as the Blue John Cavern near Castleton in Derbyshire. This is a natural cavity worked as a mine for the sake of obtaining the elegant fluorspar which gives its name to the site (the miner's name for this stone being Blue John), and which is here found in small detached pieces in the limestone rock. "The principal subterranean apartment is termed the 'Hall,' a wide and lofty cavern such as imagination conceives would be a fitting home for the romantic outlaws of Salvator Rosa. The perpetual waters which trickle down have left a residuum of lime, which has been moulded by accident and by industrious and gentle operation into a thousand free tresses and waving bands. The whole is fashioned by nature with less of the abrupt form which characterises the congelation of fountain streams by cold, and presents a grotesque enamel of exquisite polish and gracefulness, giving to the artificial plain or coloured lights, uplifted within the conical abyss, beautiful reflections from its unrivalled crystallised surfaces." [1]

English cavern scenery.

The temperature of caverns is generally very constant and a little lower than the latitude would indicate, but not often falling to freezing-point, whilst very deep caverns and mines exhibit a distinctly warm temperature. The cave of Schaflock in the high Alps presents scenery of a striking kind, in which ice effects are conspicuous. It is bored in the solid rock, "a vast castellated wall," nearly 1000 feet above the level of the valley, and is extremely difficult of access. [2] A traveller thus describes this cavern: "Now opened upon us in the deep gloom a splendid scene. Not many feet beyond us blazed innumerable stars, which glistened like sparkling diamonds in the ebon horizon. Wherever the light of our lamps fell, a rainbow radiance illumined a little sphere, which twinkled and sparkled

[1] Milner's *Gallery of Nature.*
[2] *Temple Bar Magazine*, vol. iii.

like a planet on a dark December night. From the
roof to the ground this brilliant galaxy of stars
continued in a broken, but not less beautiful chain.
Place a piece of phosphorus in a rayless gloom, or the
pretty coruscations of the glow-worm on a sylvan bank

FIG. 31.—Logan Stone near Land's End.

in June; magnify the effect a thousandfold, and then
you may conceive something of the witchery of the
picture of which we were witnesses." There were
stalactites and filigree work, all of ice, imitative of
kiosks, minarets, pavilions, and little temples whose
architecture was the handiwork of water and ice.

G

The study of scenery accessible.

The scenery of a country thus presents a constant and inexhaustible field of study. If many of nature's choicest effects are associated with other lands than ours, it is nevertheless true, as set forth by Sir Archibald Geikie,[1] that "the surface of every country is like a palimpsest which has been written over again and again in different centuries, and how it has come to be what it is cannot be told without much patient effort. But every effort that brings us better acquainted with the story of the ground beneath our feet, and, at the same time, gives an added zest to our enjoyment of the scenery of the surface, is surely worthy to be made." The denudation and sea work upon the cliffs upon the west coast of Ireland; the destruction by the sea of the land now represented by the Needles off the Isle of Wight; the curious weathering action upon the Cornish coast (Fig. 31), and the geological changes that have occurred in the neighbourhood of Loch Maree are interesting examples; but in every part of the world there are abundant resources for the studious, for there is no landscape "which may not be examined with fresh interest if the light of scientific discovery be allowed to fall upon it. Bearing the light with us in our wanderings, whether at home or abroad, we are gifted, as it were, with an added sense, and increased power of gathering some of the purest enjoyment which the face of nature can yield."

[1] In the *Fortnightly Review*, April 1893.

IV

THE SEA

"Thou art sounding on, thou mighty sea, for ever and the same ;
The ancient rocks yet ring to thee, whose thunders naught can
    tame.
The Dorian flute that sighed of yore along thy waves is still ;
The harp of Judah peals no more on Zion's awful hill.
And Memnon's too hath lost the chord that breathed the mystic
    tone,
And songs at Rome's high triumphs poured are with her eagles
    flown ;
And mute the Moorish horn that rang o'er stream and mountain
    free,
And hymns the learned crusaders sang have died on Galilee.   ·
But thou art swelling on, thou deep, through many an olden
    clime,
Thy billowy anthem ne'er to sleep until the close of time."

Felicia Hemans.

OF all the earth's features none will compare in im-
pressiveness with the ocean.   Its immense expanse, its
incessant movements, its varied and striking phenomena,
the scenery of the rocks and islets scattered over its
surface (Fig. 32), the multitude of its inhabitants, and
the richness of its harvest, its permanence and manifest   The place
power, combine to give the sea the foremost place in   of the sea
the scheme of nature.   "Perhaps," says Mr. Gosse the   in nature.
naturalist, "there is no earthly object, not even the
cloud-cleaving mountains of an Alpine country, so
sublime as the sea in its severe and naked simplicity."

In the Scriptures are numerous telling allusions to the sea, to which some of the early classical writers are also indebted for some of their imagery. Homer, the father of poetry, sang of the sea and its wonders; and Virgil, with facile touch, exerted his powers in its description. By turns it is deep, mighty, vast, moaning and foaming; but it is also restless and faithless. The Romans, the Greeks, and the Hebrews were in fact impressed by the force and stormy character of the sea rather

Fig. 32.—Æolian Rocks, Bermuda.

than by the beauty of its aspects, and there is manifest a feeling bordering upon aversion in some of the references to it. In the book of Job we read of its "proud" waves; and a modern writer, speaking of this relationship, remarks: "The ocean, the sky, the weather were too fatal for man to be lightly dealt with in the stark reality without much investigation of details, always beautiful when sufficiently seen." To the Romans the sea was malevolent rather than beneficent, for in those

Early ideas.

days nature was regarded as mysterious and oppressive, and the ocean in its boundlessness and masterful-
ness filled a leading place. In the Mediterranean the
Phœnicians, and in the Northern Seas afterwards the
Vikings, acquired command of the sea ; and, losing all
dread, began to entertain for it a feeling of admiration,
which has been handed down to the present times. Man's control of the
sea.

Fig. 33.—Ruins of a Viking Ship.

The Vikings were often arrested in their marauding
expeditions by the storm winds carrying them the way
they would not go ; but if the tempests overcame their
skill and destroyed their fantastic-looking ships (Fig. 33)
no lamentation was heard ; rather, with King Olaf,
they were ever ready to sing :—

" For of all the runes and rhymes
Of all times,
Best I like the Ocean's dirges,
When the old harper heaves and rocks
His hoary locks,
Flowing and rushing in the surges."

The modern mastery of the sea by the English is
exemplified by the fact that over half the world's mer- English
love of the
sea.

cantile marine is owned by Great Britain; the love of
exploring it is shown still by our sea captains, and
yachting is as popular as ever.   We rejoice in our sea-
girt home, around which the waves are no less tempestuous than when they baffled the Roman invaders, and,
long afterwards, the Spanish Armada; but the appliances
of science give a control previously unknown.   The
ship captain need not now seek for port in a storm,
but, trusting to the command that steam gives him,

FIG. 34.—Macleod's Maidens and the Basalt Cliffs of the West of Skye.

may even put out to sea and face the gale, as did
the Channel Fleet in a heavy storm a few years ago.
The far heavier attacks of tropical tempests have
also been withstood, as when, off the island of Samoa,
H.M.S. *Calliope*, opposing ten thousand horse-powers
to the force of wind and wave, rode successfully
through — an act of seamanship that will long be
remembered.

Its extent.   The ocean occupies $\frac{7}{10}$ of the earth's surface, but
forms only $\frac{1}{1786}$ of its mass, so that, vast as is its
extent, and deep as are its waters, it is yet but a super-

FIG. 35.—View on the East Coast of Scotland.

ficial coating of the great globe itself.   The voyager is,
however, impressed by its enormous area.   It has been
computed that to stand in a fresh square mile of land
each minute would occupy a hundred years, but for a
similar survey of the ocean's broad expanse nearly three
hundred years would be required.

The depths of the sea have been determined with
accuracy only in quite recent times.   In the early part
of the present century a ship captain would occasionally
lower a cannon-shot to gauge the depth of water under
him if in the shallows; but such a method would be
misleading if practicable in water of 1000 fathoms.
Deep-sea sounding has now been reduced to a system,
English and American navigators being emulous of
obtaining the best results; and, by the use of special
apparatus, the depth of the sea has been correctly
ascertained, and important facts regarding the ocean
floor been revealed.

The sea depths. There is found to be a correspondence between the
height of the land and the depth of the sea.   The loftiest
mountains are four or five miles above the sea, and,
where deepest, the ocean floor is as far below the surface.
Off Newfoundland, in the North Atlantic, five miles
have been registered; and near Japan in the North
Pacific, near the Caroline Islands in the South Pacific,
and in the vicinity of the West Indies, equally deep
soundings have been obtained.   Of the whole ocean
bed eighty per cent is from 6000 to 18,000 feet, or,
roundly, from 1 mile to $3\frac{1}{2}$ miles below the surface.

Our narrow shallow seas. Around the British shores the sea is shallow.   In
the German Ocean our tallest factory chimneys would, in
many parts, show above the surface if planted upon the
bottom; and, in the short run from Liverpool to the Isle
of Man, objects upon the bed of St. George's Channel can
in very calm weather be seen for much of the distance.
Off the Scottish coasts (Figs. 34, 35) a depth of 300
or 400 feet is reached.   But, some 200 miles to the
west of Ireland, there is a dip to over a mile

where the true floor of the Atlantic begins.   The

Fig. 36.—"The Old Man of Hoy," Orkney Islands.

weather around the British coasts is frequently stormy,

and the winds and waves have carved the rocks into
fantastic and picturesque outlines (Figs. 36, 37, 38, 39,
and 44). The systematic sounding of the sea, in
which the *Challenger* steamship some years ago took so
important a part, has led to the discovery of extended
levels in the Atlantic, upon which the telegraphic cables
can be laid securely. From Cape Race to Cape Clear,
a distance of 1640 miles, such a level exists. These
levels have been named "telegraphic plateaus," from
their importance in ocean telegraphy. In some parts,
far away from land, shallows have been discovered
which may afford convenient anchorage, and the unusual sight be seen of a vessel riding at anchor in mid-
ocean.

The ocean floor. The bed of the sea is not composed of sand, as is
so often supposed, for sand is a formation of the shore.
Far out in the ocean the sounding apparatus brings up
either a kind of gray or bluish-white mud, composed,
as the microscope shows, of myriads of shells or cases
of minute marine creatures of the lowliest type of
structure, spoken of as "globigerina ooze," from being
composed of globe-shaped foraminifera, and which ooze
occupies fifty millions of square miles of the ocean floor.
At lesser depths the character of the ooze changes with
the varying character of the foraminifera composing it,
some ten millions of square miles, for example, being
occupied with ooze composed of diatoms. Certain
areas have a flooring of red clay, which was long a
puzzle to the scientists, as clay, like sand, is generally
a product of the shore or of the land, and not of the
deep sea. The explanation is that this clay is not a
direct but a secondary deposit, due chiefly to the disintegration of pumice stone and other igneous rocks,
which are extensively found in volcanic areas, in which
areas the red clay abounds. The dredge often collects
the teeth and other imperishable remains of sea creatures, a single haul sometimes bringing up a thousand
sharks' teeth, mixed with the ear bones of whales,

FIG. 37.—The Land's End, showing the wasting action of the sea.

affording evidence of life in deep waters in times that
have passed away.

Great pressure at great depths. In close connection with depth is that of pressure.
Water, like air, exerts its pressure in a very even
manner, but the effect is not less on that account. At
great depths it becomes enormous, amounting roundly
to a ton per square inch one mile deep, or to three
tons per square inch at three miles deep. How the
humbler creatures of the sea brought up from such
depths can sustain this pressure without injury is
Hydrostatic paradox. explained upon the principle of the hydrostatic paradox,
a peculiar law of fluid pressure under which a body of
water infinitely large is held in balance by a quantity
extremely small. The action is akin to that of the
atmosphere upon the human body. Man lives at the
bottom of the aerial ocean, his element, which reaches
overhead to the height of more than a hundred miles,
exerting upon him a pressure varying within small
limits, but averaging fourteen pounds to the square
inch, and amounting, upon the surface of the body of a
man of ordinary size, to fourteen tons. Yet he bears
this load easily — unconsciously, in fact — the small
amount of air within his body forming a perfect
counterpoise to the whole outside atmosphere. If the
equilibrium be disturbed, as occurs in cyclonic storms
(it may easily, in a mimic way, be imitated by experiment), the effect is striking, either in air [1] or water. In
whale fishing it has occasionally happened that a boat
becoming entangled in the harpoon line has been drawn
down by the rush of the wounded whale to such a
depth as to be useless from being compressed and
waterlogged.

The composition of sea water is explained by its

[1] The equal pressure of the air may be thus disturbed. Experiment : Obtain a wine-glass, into which place a little thin dried paper.
If the paper be lighted and allowed to burn for a time, and then be
closed by the hand, the paper will burn for a short time, using up a
small portion of the air. The glass will now be firmly pressed to the
hand by the outer air being to a small extent unbalanced.

FIG. 38.—The Steeple Rock, Kynance Cove, Cornwall.

origin and early history. The solids, chiefly common salt, contained in sea water are in much greater quantity than is ordinarily supposed, a misconception fostered by its transparency. The ocean brine contains about three and a half per cent of solid matter, or over two pounds in one cubic foot of water. If the sea could be evaporated, leaving the solid residue, there would be a layer, chiefly of salt, over the whole bed of 350 feet, say nearly the height of St. Paul's Cathedral; or, if the land surface only could be overspread, the solid matter might be piled up everywhere to the height of 1000 feet. The salt harvest is illustrated by Fig. 40.

The salt-ness of the sea.

Sea water contains nearly every known chemical element, and the term "brine" would be more descriptive of its character.[1] The sea is saltest where deepest, and where far removed from the presence of rivers. The salinity is high between 22° north and 17° south latitude, the region of the trade winds, where evaporation, caused by constant winds and the sun, is unusually active. Ice reduces the salinity. The Mediterranean is uncommonly salt, owing to copious evaporation and the few rivers entering it. Around the English coasts sea water has a specific gravity of 1027 (distilled water being taken at 1000). The Baltic, having many rivers, besides being shallow, is less salt, 1015. Solids in sea water, by increasing its buoyancy, assist navigation. A ship loading at London or any river port would be more buoyant when gaining the open sea. Other solids than salt contained in sea water are the chloride of magnesium, sulphate of

Varying weight of sea water.

[1] Percentage composition of salts in Atlantic water :—

Chloride of magnesium	.	.	.	0·360
Chloride of sodium (common salt)	.	.	.	2·700
Chloride of potassium	.	.	.	0·070
Sulphate of lime	.	.	.	0·140
Sulphate of magnesia	.	.	.	0·231
Bromide of magnesium	.	.	.	0·002
Carbonate of lime	.	.	.	0·003
				3·506

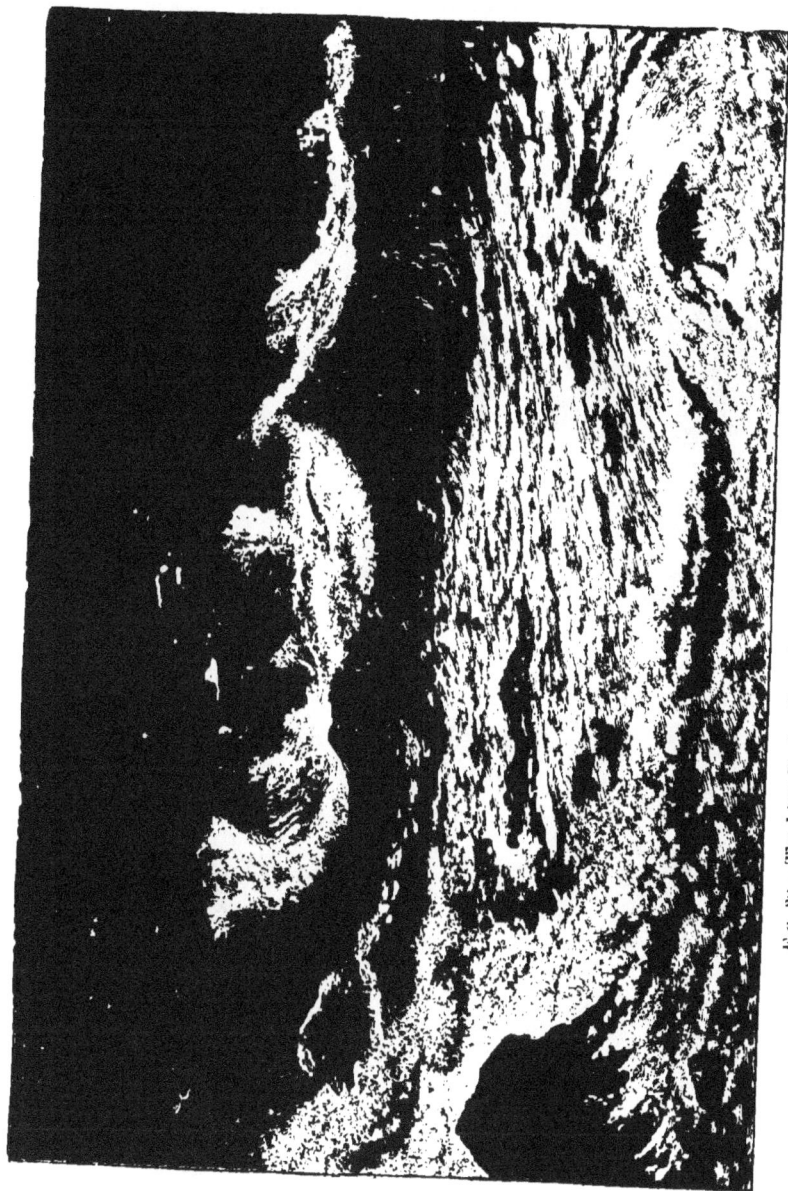

Fig. 39.—The Lion Rocks, Cornish Coast—a sea in which nothing can live.

soda, and carbonate of lime and silica.[1] The carbonate

FIG. 40.—Salt Pits in the West of France.

of lime is used up by the mollusca constructing their

[1] It is instructive to precipitate the chlorides in a test-tube containing sea water by pouring into it a small quantity of nitrate of silver. The hydrometer may also be used to illustrate the increased weight of sea water, or a very light fishing float will suffice, placed first in fresh water and afterwards in sea water.

shells of it, and the coral animalculæ their polypidom, which is often of enormous dimensions.

The colour varies from a variety of causes. Around the British shores the shade is blue - green and the waters are pretty transparent, as is also the case

Fig. 41.—View in a Norwegian Fjord.

in the Arctic Ocean. Amongst the West Indian Islands, and more notably in the Scandinavian fjords, the water itself becomes almost invisible (Fig. 41). The clearness of the northern seas is remarkable. "Hanging over the gunwale of the boat," says a traveller, "with wonder and delight I gazed on the slowly-moving scene below. Where the bottom was sandy, the different

H

Trans-
parency.

kinds of asteriae, echini, and even the smallest shells appeared at that great depth conspicuous to the eye. Now creeping along, we saw, far beneath, the rugged sides of a mountain rising towards our boat, the base of which, perhaps, was hidden in the great deep below. As we pushed gently over the last point of it, it seemed almost as if we had thrown ourselves down this precipice, the illusion from the crystal clearness actually producing a sudden start."

Miss Martineau says of the fjords of Norway: "It is difficult to say whether these fjords are the most beautiful in summer or in winter. In summer they glitter with green sunshine; and purple and green shadows from the mountain and forest lie on them : and these may be more lovely than the faint light of the winter noons of these latitudes, and the snowy pictures of frozen peaks which then show themselves on the surface; but before the day is half over, out come the stars—the glorious stars which shine like nothing we have ever seen. More, the planets cast a faint shadow, as the young moon does with us : and these planets, and the constellations of the sky as they silently glide from peak to peak of these rocky passes, are imaged on the waters so clearly that a fisherman as he unmoors his boat for his evening task feels as if he were about to shoot forth his vessel into another heaven, and to cleave his way among the stars."

Virgil describes the waters of the Mediterranean as "azure purpureum." In the Indian Ocean are shades of black; in the Gulf of Genoa a white hue occasionally prevails, and the Sea of California is reddish. When the sea is shallow the bottom shade shows through and

Phosphor-
escence.

affects the colour. One of the finest effects is that of phosphorescence, an appearance often due to vast multitudes of minute luminous creatures—"glow-worms of the sea." The appearance is most striking in the tropics. "The path of a vessel," says an observer, "seems like a long line of fire, and the water thrown up in her progress or dashed by the waves upon deck flashes like lambent flame. Sometimes myriads of luminous stars float and dance upon the surface, assuming the most varied and fantastic aspects."

The temperature of the ocean is much less variable than that of the land. At the equator there is a difference of only 3° or 4° Fahr. between day and night; whilst inland the range may be considerable, after the great heat of day the night temperature in some parts falling to freezing-point. In hot seas the temperature decreases with depth. In the Caribbean the surface water is often 83° Fahr., whilst at a mile deep it is only half as much. The surface temperature ranges from 28° Fahr. to 86° Fahr., but at very great depths is pretty uniform, 32° to 38° Fahr. Except in landlocked seas, like the Mediterranean, the bottom water is cold. In the Mediterranean the temperature approximates to that of the land a few feet below the surface. The ocean is thus an equaliser of temperature. In the tropics sea winds cool the land by day and warm it at night, and so prevent those extremes of temperature which are so injurious to the human constitution. Islands and insular regions, if ruled from the sea (that is, having chiefly sea winds) thus enjoy an equable and pleasant temperature.

Temperature.

The sea is in constant movement. The tidal wave follows the apparent course of the moon towards the west, which is its normal direction, but rebounding if it encounters a mass of land in its path, as on reaching the American continent after crossing the Atlantic, when the course of the waters is almost reversed, so that high water comes to these shores from the south-west and not from the east. On the celestial mechanism of the tides and the conditions complicating the problem of tidal motion, it is not here the intention to enlarge, the object being to refer only to the external aspects and to the zest and freshness which the tidal movement gives to the shores. The tides around our own shores

Oceanic movements.

Tidal motion.

are very distinct, although not the highest that could be named. The illustration (Fig. 42) shows the effect of the varying position of the sun and moon in producing the tides.

Ocean rivers.   The sea has its rivers. Compared with them, the Amazon, the Nile, the Mississippi, and the great streams of the land are insignificant. These ocean rivers have their beds and their banks, composed of water of a

SPRING-TIDES

NEAP-TIDES
FIG. 42.

different density and temperature from the current itself. They are originated by difference of temperature of the water in the several zones, which leads to an exchange, and which is modified by the action of the winds, the distribution of the land, and the eastward rotation of the earth.

The Gulf Stream.   Each of the oceans has its currents, several of which are of the first importance as bearing upon the climate of countries near which they flow; but none will compare with the Gulf Stream of the Atlantic for its effects

upon Britain and the western countries of Europe. As
the ocean currents form closed circles it is hardly logical
to fix upon any point as their origin. The great body
of hot water known as the Gulf Stream issues from the
Gulf of Mexico, from which its name is derived (and
which, with the connected Caribbean Sea, has been named
the earth's cauldron), between the Bahama Islands and
Florida on the mainland of North America. It is here
about 60 miles wide, has a temperature of nearly
90° Fahr., a depth of a quarter of a mile, and a velocity of 4 miles an hour. Upon entering the Atlantic
it broadens out, moving along the American shores
until reaching 40° north, when it turns distinctly to
the east, here being 75 miles wide, with a decreasing depth. Proceeding across the Atlantic, it loses
speed and depth, but spreads out to over 100 miles,
very slowly losing the high temperature which it acquired in crossing the Atlantic under almost vertical
sunshine, and afterwards when cooped up in the
"cauldron" referred to, travelling over 10° of latitude
for the loss of only 6° of temperature. The importance
of this ocean river, the portion of which we are concerned
with being over 3000 miles long, can hardly be over-
estimated when the specific heat of water is considered
in relation to that of the winds. The stream proceeds
expanding like a fan, and carrying the moisture and the
heat of the tropics into the lands of the temperate zone. A great
When nearing the British shores the prevailing south- ocean river.
westerly winds, which also help forward the stream, carry
its influences over the land, and the English climate in
its humidity and its even temperature, with its cool
summers and mild winters, is principally due to this
ocean river. There are some drawbacks. Sailors speak
of the Gulf Stream as the "storm-raiser" or the "storm-
breeder," and the tempestuous character of the North
Atlantic is mainly a consequence of its action.

The Japanese current, "the Gulf Stream of the The Gulf
Pacific," leaving the East Indies, travels south-west Stream of
the Pacific.

along the coast of Australia, and north-west along the
eastern shores of Asia. Its elements have not been
thoroughly ascertained. It is of so dark blue a colour
as to have earned the designation of " the black stream,"
and is unusually salt. It carries the driftwood contributed by the Chinese rivers to the Aleutian Islands,
which are treeless, and where the timber thus carried is
of the utmost value, the people depending upon the
ocean supply for the construction of their boats, buildings, and furniture.

Wave motion. The wave motion of the sea has been carefully
observed, especially in the Atlantic. Waves are originated by the wind, by submarine volcanic action, and by
the tides. Waves caused by the winds are superficial,
are a movement of form only, and move at right angles
to the direction of the wind. The wave motion is
rapidly enfeebled from the surface, the largest waves
having no influence at a distance below greater than
their breadth. This may be as much as 500 feet.
Popular ideas regarding waves are very erroneous.
Their height has been greatly exaggerated. There are
no waves " mountains high," or even approaching the
elevation of a tall building. Around the British
shores (Fig. 43) the distance from crest to trough of
a wave rarely exceeds 8 or 9 feet, whilst the greatest
waves measured—those formed off the Cape of Good
Hope in the "roaring forties"— are sometimes 30
or 40 feet. But the quantity of water in a wave
and its weight and destructive power are under-estimated by landsmen. Captain Scoresby, under many
difficulties, made careful calculations of the dimensions
of Atlantic waves, and concluded that the altitude
of the greatest wind waves in this ocean is 43 feet;
that in a heavy storm the distance between the waves
may reach over 500 feet, and that the velocity of a
wave's progress may be over 30 miles an hour. It
would be thus easy to show that an Atlantic roller
would contain hundreds of tons of water, and, when it

FIG. 48.—Marine View on the West Coast of Scotland.

is borne in mind that the motion is occasionally as swift as a railway train travelling express, an idea may be formed of the destructive action when a ship is "struck by a sea."

Islands. Water-lands.

Islands (etymologically water-lands) belong to the sea, at any rate those of the pelagic type situated far from land. They have been called "the brooches of the sea," and well deserve the epithet from their attractiveness. Their origin has been chiefly of a twofold character. Some groups, as the Azores in the Atlantic, and the Sandwich Islands in the Pacific, are of igneous formation; they have been raised by the elevatory power of the earth-fires existing beneath the sea-bed Within the historic period new islets have been raised from the deep.

Other groups are due to the industry of the coral polype. The first step is to begin the polypidom around the flanks of an island at a depth not exceeding thirty fathoms, below which they cannot carry on their

Coral islands.

reef-building. Many of these islands rise from parts of the ocean floor that are sinking slowly, and in this case the coral polypidom would as slowly rise, keeping level with the sea. At length the island would disappear, and an annular-shaped coral island or atoll would mark the place of its descent.

Maintained at the sea-level, the polypidom grows laterally. Winds and currents deposit upon it the debris of the sea; birds and sea creatures visit it; and, presently, there is a humble growth of vegetation; plants higher in the scale of life afterwards appear, *e.g.* the cocoa-nut palm; and at last, when the atoll or coral island is completely furnished, man takes possession of it. The island thus formed presents a steep face to the sea all round, with the exception of a break where a current has entered, sometimes descending to the depth of 1000 feet, and forming a structure, compared with which the most Titanic masonry of man appears insignificant.

Coral islands are formed in tropical seas, limited to about latitude 28° north and south, and between 134° west longitude and 135° east. The Maldives and the Laccadives, off the coast of India, Mauritius, and the Bermudas are good examples of coralline islands.

The uses of islands are many and important. Some- *Uses of* times they stand upon trade routes, as the Mauritius *islands.* and the Falkland Islands; others are held for imperial reasons, as Malta, from which British interests in the south of Europe may be watched. Others are useful to certain industries, as Newfoundland, which forms a base of operations for the cod fishery. Others are favoured health resorts, as the Isle of Man, which is accessible from the great industrial centres of Lancashire and Yorkshire; and the Isle of Wight, which is easily reached from the southern counties. Occasionally an outlying island has served as a state prison, as St. Helena for Bonaparte, situated as it is thousands of miles away from European political influences.

The Pacific Ocean is of compact form, extending *The oceans.* over 145° of longitude, nearly half round the world, and is an immense sunken basin, with its margins rising into mountains. Its area is nearly equal to the whole extent of the land, and its island groups are its most characteristic feature. The seas connected with it are not of the first importance.

The Atlantic, although only half the size of the Pacific, is of higher commercial use. Upon it are the "steam lanes" traversed by ships connecting the Old World with the New. This ocean, more than any, is a "commercial field for the union of distant peoples." Its coasts are long, forming bays and harbours, and so further facilitating marine enterprise. It has been computed that, although so far inferior to the Pacific in area, it would take a ship longer to explore its coasts—some eleven months, at the rate of a thousand miles a week, which is more by about six weeks than would be occupied at the same rate upon

the Pacific. Besides its smaller articulations, there are
several independent parts or seas with special names,
which have played an important part in the growth of
civilisation. On the American side are the Caribbean
Sea, the Gulf of Mexico, and Hudson's Bay ; on the
European side the Baltic and the Mediterranean.

The Medi-
terranean.

The last-named is the most historic of all seas, and
its physical habitudes are also of the greatest interest.
Its area is not less than a million of square miles. Its
coast-line is highly articulated, and its northern shores
"one girdle of perpetual beauty." Its waters are salt,
deep, buoyant, and of an indigo-blue colour, whilst the
flora of the countries bordering upon it is extremely
rich. The feelings of a scholar and student of nature
upon beholding this sea for the first time are thus
expressed : "There it is at last, the long line of
heavenly blue, and over it, far away, the white-peaked
lateen sails, which we have seen in pictures since our
childhood, and there, close to the rail, beyond the
sandhills, delicate wavelets are breaking for ever on a
yellow beach, each in exactly the same place as the one
which fell before. One glance shows us children of the
Atlantic that we are on a tideless sea. There it is, the
sacred sea, the sea of all civilisation, and almost all
history, girded by the fairest countries in the world,
set there that human beings from all its shores
might mingle with each other and become humane.
The sea of Egypt, of Palestine, of Greece, of Italy,
of Byzant, of Marseilles ; and this Narbonnaise,
more Roman than Rome herself, to whom we owe the
greatest part of our own progress ; the sea, too, of
Algeria, and Cyrene, and fair lands now desolate, surely
not to be desolate for ever. Not only to the Christian or to the classic scholar, but to every man to
whom the progress of his race from barbarism
toward humanity is dear, should the Mediterranean
Sea become one of the most august and precious
objects on the globe ; and the first sight of it should

inspire reverence and delight, as of coming home—
home to a rich inheritance in which he has long
believed by hearsay, but which he sees at last with his
own mortal eyes." "The air," says the same eloquent
writer, "is glassy clear as the water, and through it, at
seemingly immense distances, the land shows purple
and orange, blue and gray ; and all the landscape is
one great rainbow."

The Indian Ocean is triangular in shape, and has an Indian
extent nearly equal to the Atlantic. It was formerly Ocean.
crossed by ships trading with our great eastern de-
pendency, India ; but the opening of the Suez Canal
has shortened the journey over it.

Of the circumpolar oceans the Antarctic is little The Polar
known. It has no land boundaries, and its navigation seas.
is very difficult and dangerous. In the Arctic Ocean,
however, Asia, Europe, and America are closely interested, and several expeditions have essayed to determine its geography and characteristics. Its diameter
is some 2400 miles, and there is an impression that its
centre is an open ocean, that the coldest climate and
greatest ice barriers are at the circumference. The
latest attempt at its exploration was referred to in a
previous chapter of this series, and geographers and
scientific men will await the return of Dr. Nansen's
expedition with much interest. The proof of a navigable sea around the North Pole would be of much significance to the mercantile world. Several of the journeys
from ports of Northern Russia would be immensely
shortened could a polar route be found, if only open
part of the year, and England and America would also
benefit. A glance at the terrestrial globe will show
that the most direct journey from China to England
would be northwards, and the carrying trade of the
whole northern hemisphere might be influenced by
such a discovery ; but there would be great physical
obstacles in the way of a regular polar route. The
navigation of the Arctic seas as at present known is

full of difficulty. Icebergs and ice-fields encumber the way, and intense cold renders many ordinary operations impossible, whilst the alternations of light and darkness are in these latitudes extremely inconvenient. The scenery is often striking and impressive, the icebergs, in their massiveness, their picturesque — sometimes grotesque — outline, displaying fine effects of colour.

Of the inhabitants of the sea, shell-fish abound everywhere, some of them so small as to be microscopic, but often of extraordinary beauty of outline and colour. In the days when conchology was popular and shell-collecting a favourite diversion, high prices were paid for rare specimens, sometimes £20 or £30 being given for a specimen of the Chinese wentle-trap (*scalaria pretiosa*). Nothing in nature is more exquisite than some of these shells, in form, in their minute sculpturing, and their rich hues. Tennyson expresses the popular admiration of these products of the sea :—

"See what a lovely shell,
Small and pure as a pearl,
Lying close to my foot,
Frail, but a work divine.
Made so fairly well
With delicate spire and whorl,
How exquisitely minute—
A miracle of design.

.   .   .   .   .

"The tiny cell is forlorn,
Void of the little living will
That made it stir on the shore.
Did he stand at the diamond door
Of his house in a rainbow frill ?
Did he push, when he was uncurled,
A golden foot, or a fairy horn,
Through his dim water-world ?"

But the sea has not only treasures for the cabinet. Many of the mollusca furnish excellent food. The

oyster, the whelk, the cockle, the mussel, and the periwinkle minister to the appetites of the rich and the
poor. The periwinkle, for example, is brought in tons
to Billingsgate market, and the flavour of these sea   Harvest of
snails is said to be "delicate and relishing." The   the sea.
general fish supplies it would be impossible here to
describe in any detail. Besides the mollusca there are
the crustacea—crabs, lobsters, and shrimps. In the
British seas the herring is caught chiefly on the east
coast, Yarmouth, Lowestoft, and Grimsby being especially engaged in this fishery. Cod-fish is taken on the
Yorkshire coast, and the pilchard in countless numbers
off Cornwall and Devon.

Amongst the miscellaneous treasures of the sea may
be mentioned sponges, pearls, and coral, the obtaining
of each of which forms a distinct industry. The sea
has thus its harvests like the land, sometimes seasonal
and periodic, but oftener constant and conditioned only
by the weather.[1]

Little more than an enumeration can be given of
the varied uses of the sea. It is the storehouse of
vapour and rain. There is a continual exchange   Uses of the
between the waters of the land and the sea. The rain   sea.
produces rivers, which run into the sea, and, becoming
evaporated, pass over the land and are again condensed
into rain, the ocean thus forming an important link in
a circle of endless change. The sea makes its mark in
the humidity of the countries bordering upon it, and
the luxuriance of tropical vegetation is a direct consequence of the abundant moisture from the sea. Then
the ocean is the "highway of the nations." The waters
within three miles of shore belong to the various
countries, but beyond that limit they are cosmopolitan
and free ; so that the sea, whilst physically separating

---

[1] As showing the value of some of the minor industries of the sea,
the almost unknown island of Bahrein in the Persian Gulf may be
named as having a pearl fishery of the annual value of a quarter of a
million sterling.

the peoples of the earth, really binds them together with a girdle of commerce and goodwill. Water carriage is now nearly as rapid as transit by land. The latest "greyhounds" of the Atlantic are timed to cross in about five days ; and whilst conveyance is swift it is economical also, as compared with the railway service, and commodities can thus be exchanged by sea with the greatest advantage.

That the sea moderates climate has been touched upon in connection with temperature. On a large scale the ocean has an equalising influence, moderating the fierce heat. of tropical lands, and tempering the rigour of polar realms ; and so this earth is made a pleasanter abode.

In the early time, it has been shown, man was impressed with the sterner aspect of the ocean ; it was considered only "in a terrible point of view without reflecting on the wonders and blessings it so visibly presents." But, in these later days, having tamed its power and reduced it to servitude in the interests of mankind, we may admire its varying moods, whether in storm or calm, be revived by the vigour of its unending motion, and elevated by the contemplation of its vastness. A modern writer observes : "In the wide sphere of bright creation there exists nought that hath for men so deep a tone of meaning as the fathomless eternal sea — that resplendent shield guarding the verdant universe. It hath smiles for him in his gladness, when the glorious sun, dancing over the tameless waves, lights them into beauty ; it hath a garb of mourning for his sorrow, when it reflects the dark cloud sailing over it and rocks the shadow within its bosom ; it hath notes of laughter for his hour of wassail and of song, when its free bright waters leap to shore with a sound of bounding mirth ; and it hath a trumpet for the victor when it raises its voice amidst the storm, and sends its billows gleaming on high like mighty standards. Thou hast within thy depths, O sea,

FIG. 44.—The Stacks of Duncansby, Caithness, a wave-beaten coast-line.

·gems to deck the brow of the beautiful, wealth to lure the aspirations of the avaricious, and groves of rich red coral to haunt the poet's dreams. Thou hast thy treasures amongst the dead, to fill the soul of the mourner. Thou art, O sea, the deep heart of earth, imaging its beauties, thoughts, and passions."

V

RIVERS AND THEIR WORK

> "The lapse of time and rivers is the same,
> Both speed their journey with a restless stream ;
> The silent pace with which they steal away,
> No wealth can bribe, no prayers persuade to stay ;
> Alike irrevocable both when past,
> And a wide ocean swallows both at last.
> Though each resembles each in every part,
> A difference strikes at length the musing heart :
> Streams never flow in vain ; where streams abound,
> How laughs the land with various plenty crown'd ;
> But time, that should enrich the nobler mind,
> Neglected, leaves a dreary waste behind."

THE ancients, as previously stated, regarded the sea in its magnitude and power as oppressive and malevolent, as seriously entering into conflict with man in his passage through this world. In the Apocalypse, in the vision of the new heaven and the new earth, "there was no more sea"; but "a pure river of water *Rivers* of life, clear as crystal, proceeding out of the throne of *symbols* God." Milton seems to have had such a river in view— *of bene-* *ficence.*

> "The river of bliss through midst of heaven
> Rolls o'er Elysian flowers her amber stream."

And in the Scriptures the references to the influences of rivers are to their beneficence. The author of the *Ancient and Modern History of the Rivers of the Bible*

I

remarks : "In such lands nothing is more natural than that fresh streams and flowing rivers should be constantly used by the inspired poets and prophets as emblems by which to call up and shadow forth the sweetest hopes and most comfortable thoughts. Is peace spoken of? 'It shall flow as a river.' Is judgment or righteousness prayed for? It is that it may 'run down as waters, and as a mighty stream.' The advantages of wisdom in a man's heart are as 'deep waters' and as 'flowing brooks,' and the divine protective care is symbolised by the 'river the streams whereof shall make glad the city of God.'" In the lands of the Bible, where the climate is hot and rain infrequent, the streams carrying clear, cooling water from the heights refreshed the land in a degree unknown to the dwellers in Western Europe. The Tigris and the Euphrates were rivers earliest made known, and they dominated the district through which they flowed. Of the Euphrates a traveller states :

The Euphrates.

"The banks of the river and the whole extent of the valley are like a vast flower garden. Beds of poppies, scarlet and white ; bugloss and larkspur of the richest azure ; white blushing cistuses ; anemones with white, scarlet, and delicately-pencilled petals ; ranunculuses,

That great river.

and campanulas, and a thousand other flowers, with names unknown to us, display their beauties or diffuse their fragrance on every side. Chiefly bulbous-rooted plants abound in this region : wild tulips, white, red, and blue ; yellow daffodils, jonquils, gladiolus ; hyacinths of many species ; cyclamens with drooping blushing blossoms, and lilies of every gay hue shoot up their sword-like leaves, and expand their lovely corollas from the mossy turf, enamelling its surface like a gorgeous carpet." The Euphrates valley, naturally rich in soil, was rendered highly productive by numerous irrigating canals and hydraulic engines, the yield of corn being sometimes two hundredfold. In a lesser degree the river Jordan, and the "water

brooks" of Palestine, kept continually before the
inhabitants of that favoured land the helpful work
of rivers.

But it was only when the fine proportions and
regularity of the flooding of the Nile came to be
realised that it was seen that a river may be one of

FIG. 45.—On the Nile : a Cargo Boat.

nature's greatest workers for the advantage of mankind. Pliny truly described Egypt as "the gift of the
Nile," for Lower Egypt and the delta are wholly built
up of mud carried from the highlands of Abyssinia,
the regular flooding of the river having transformed a
dreary desert into a fertile valley ; and its annual
inundations still renew the largess. In its ordinary
flow the waters are pure, the river having no tribu-

taries in its middle and lower course to carry drainage matter into it. There was a tradition that if Mahomet had tasted its waters he would have sought for im-

The Nile yields drinkable water.

mortality in order to have enjoyed them for ever. In the Soudan campaigns the waters of this fine river (Fig. 45) proved a prime necessity to the British army. On one memorable occasion, the ordinarily undemonstrative English soldiers, after a long march through the desert, in which they suffered torture from thirst, were moved to tears on sighting the Nile.

The river is in flood from June to October. A rise of about thirty-eight feet is considered the best for the country; and, in the season, the movement of the water is regularly reported, as at Cairo, where a Nilometer is placed in public view.

Rivers sometimes a national trouble.

But rivers may also be a national trouble and menace, a characteristic revealed later as new lands were discovered. The flooding of the Mississippi not infrequently brings calamity to the states it traverses in its lower course, and the Yellow River, on account of its devastating floods, has been named "China's Sorrow."

Rivers form an important link in the cosmic chain by which the ocean and the land continually exchange their waters. "All the rivers run into the sea, yet the sea is not full: into the place from whence the rivers come thither they return again,"[1] says the sacred book. The evaporated water of the seas, rising silently and unseen into the higher parts of the atmosphere, becomes visible as clouds; and, condensed as rain, descends the earth's slopes as rills and streams, ultimately swelling into rivers, which convey the waters back to the ocean reservoir, and so on in endless permutation.

This action is full of advantage to mankind. A perfect landscape requires water; and rivers, in no ordinary measure, improve the scenery and add to the pleasure of man's occupancy of the earth. In

[1] Ecclesiastes i. 7.

Fig. 46.—The Wharfe : Bolton Woods.

England the Wye, the Dove, the Thames, the Avon, and others greatly add to the attraction of our scenery, whilst Scotland, Ireland, and Wales all possess streams of uncommon beauty. Several have been raised into distinction by the painter and the poet. The Wharfe (Fig. 46), a tributary of the Yorkshire Ouse, is favoured amongst English rivers as having secured the close study of the great master of English landscape, Turner, who was enamoured of the scenery along its banks. Here, according to Ruskin, are "those shores of Wharfe which, I believe, he never could visit without tears; nay, which for all the latter part of his life he never could speak of but his voice faltered." Even in the wet weather, which so often prevails in the valley, the abounding woods afforded vistas with studies of mist and driving rain for a painter of whom it has been said "no scene was too contracted for his love, or too vast for his ambition." In rainy weather, with the river in spate, many of its characteristic features are seen at their best; but in its ordinary flow "the coffee-coloured, but not muddy, water may be seen stealing first in mere threads between the separate pebbles of shingle, and eddying in soft, golden lines towards its central currents, flows out of amber into ebony, and glides calm and deep." In one of its reaches an island divides the waters and forms a "horned flood." After heavy rains the waters are brightened by foam-bells, contrasting in an interesting manner with the dark-coloured stream :—

"Those shores of Wharfe."

> "The foam-globes on the eddies ride,
> Thick as the schemes of human pride,
> That down life's currents drive amain,
> As frail, as frothy, and as vain."

The river cuts through rocks of several geological formations, the millstone grit being well exposed at the gorge of the Strid.

Of European streams noted for their scenery the
Rhine and the Rhone are prominent. The former,
rising on the north slope of the Alps near Mont St.
Gothard, is fed by a large glacier, so that in the driest
season its waters do not fail. All along its course
numerous affluents swell its waters, nearly three thousand having been counted in Switzerland alone. Rising
at a great elevation, its current is swift; and, as it
traverses broken ground, cataracts and cascades are
formed, of which the Falls of Schaffhausen are the
finest. Here the waters pour over a precipice, and
limestone rocks spring from them. The wooded
islands in its course, the forests through which it
passes, the vineyards, gardens, and cultivated fields on
either side the river, and the numerous picturesque
ruins, each with its legend, make up a scene that can
hardly be equalled in Europe. The Rhine, as in part
the boundary between Germany and France, was regarded with patriotic sentiments by the inhabitants of
the Fatherland. Longfellow, entering into this spirit,
thus pays tribute to its charms : " Of all the rivers of
this beautiful earth, there is none so beautiful as this.
There is hardly a league of its whole course, from its
cradle in the snowy Alps to its grave in the sands of
Holland, which boasts not its peculiar charms. If I
were a German I would be proud of it, too, and of the
clustering grapes that hang about its temples as it reels
onwards through vineyards in a triumphal march."
The Rhone is almost equally famous for its scenery.
It rises two miles above the sea-level, not far from the
birthplace of the Rhine, but upon the southern slope of
the Alpine range. Also, like the Rhine, it is glacier-
fed. In its upper course it presents the usual appearance of a mountain torrent, forming cataracts and
receiving countless tributaries. Entering Lake Geneva,
it deposits mud, forming a delta. So rapidly, reckoning geologically, is this work accomplished that Port
Valais, which eight centuries ago stood at the edge of

the lake, is now a mile and a half inland. This rate
is maintained; and, in time, the lake will be filled up,
and the landscape changed. Issuing from the lake it
is joined by the rapid Arve, which brings down the
waste of Mont Blanc—in times of flood, with such
vigour as temporarily to force back the main stream
proceeding from the lake, and reverse the direction of
The Rhone. the water-wheels along the banks. The Saône and the
Doubs afterwards unite their waters, carrying eroded
matter from the Vosges and the Jura, whilst other
streams enter from the Alps of Dauphigny. Entering
the Mediterranean, it forms another delta, which has
also grown at a rapid rate; a tower, for example,
erected close to the sea in the middle of the eighteenth
century being now three miles inland. The Rhone
thus shows how a river gradually alters the landscape.

In the New World, of rivers notable for their scenery,
the Ohio, the "belle rivière" of the French colonists,
a left-bank tributary of the Mississippi, should be
mentioned. Its clear waters, its numerous tree-clad
islands, and its woods descending to the river's brink
have been specially admired. "No stream rolls for
the same distance so uniformly and peacefully." The
country around these vast watercourses is of the most
varied description, alternately exhibiting wild rice lakes
and swamps, limestone bluffs, craggy hills, deep pine
forests, and beautiful prairies, which show an almost
perfect level, in summer covered with a luxuriant
Scenery of growth of grass and flowers. "On a calm spring morn-
the Ohio. ing, and under a bright sun, this sheet of water shines
like a mass of burnished silver, its edges being distinctly
marked by a magnificent outline of cotton-wood trees,
at this time of the year of the brightest verdure, among
which those brilliant birds of the country, the black
and red bird, and the blue jay, flit to and fro, or wheel
their flight over them, forming a scene which has all
the grandeur or beauty that nature can furnish, to
soothe or enrapture the beholder." The Susquehanna

Fig. 47.—The Banks of the Susquehanna.

is another American stream of remarkable beauty
(Fig. 47).

The usefulness of rivers is indicated by the fact that
the most important cities have been established upon
their banks : in the earliest times Babylon and Nineveh
upon the Euphrates and Tigris respectively ; and
Thebes, "Homer's city of a hundred gates," upon the

Large
cities built
upon
rivers.

Nile. In more modern times London has been indebted
for its wealth and magnitude largely to its being a
river port ; and, similarly, Paris has been assisted
by the Seine, Vienna by the Danube, St. Peters-
burg by the Neva, and Calcutta by the Ganges.
In America important towns are often placed at the
junction of two streams, so as to have water communica-

Internal
communi-
cation by
rivers.

tion in three directions. The cheap and convenient exchange of commodities throughout a country is possible
in a well-watered country, not only by the rivers themselves, but also by the canals which they feed, and which
unite all the rivers of a country in a network of navigable
waters, nowhere better seen than in England, where
the internal communication is so good that prices rule
nearly the same everywhere.

Another service of rivers is in irrigation and drainage.
Where the land is but little above the watercourse,
irrigation is easy by means of artificial cuttings. In a
country like England this is unnecessary ; but, where
the rains are only periodic, and long droughts occur,
this influence of rivers is appreciated to its full extent.
In Egypt, for example, rain seldom falls—only once or
twice in the year — and notwithstanding the heavy
night dews, cultivation is successful only within reach
of the river. A modern traveller[1] has described how
the waters are led by canals from the river, one
ancient canal being over 200 miles long. These waterways were allowed to fall into decay, but steps have
been taken to restore them ; and this effort is being
vigorously followed up during the English occupa-

[1] Vide *Rob Roy on the Jordan, Nile, Red Sea, and Gennesareth.*

tion. The gracious waters are conducted quietly,
so as not to injure the banks, and are then brought
to rest over the fields, that they may deposit the
suspended mud. For half the year the waters are
obtained by the *shadoof*—leathern basins let down
into the water by hand, by balanced levers, or by
simple hydraulic contrivances. In some of the canalettes
the sluice gates are moved by the foot, which explains
the allusion in the Bible: "For the land whither thou
goest in to possess it, is not as the land of Egypt from
whence ye came out, where thou sowedst thy seed, and
wateredst it with thy foot, as a garden of herbs."[1] Under
proper management, valuable crops of cotton and maize
are raised. Six crops of clover have been gathered
in the year. Thus "the Nile is everything to the
Egyptians." A traveller, who has gauged its operations, exclaims: "Marvellous Nile, how far you spring
from, how long you wander, how many millions take
water from you: and no wonder you were worshipped
as a god." In the delta of the Nile are extensive
regions, as the "Fields of Zoan," which are not now
cultivated; but as they are annually enriched by the
flooding of the river, they have not "the sombre hue
of wild bleak savagery, but that of a rich and mellow
land."

Closely associated with irrigation is drainage, which
is also assisted by the presence of a river, although in
the time that is to come a fine river will, let us hope,
be preserved for nobler uses than that of a common
sewer, and the waste products will be utilised instead
of lost. As carrying sewage rivers may, therefore, be
expected to be a diminishing quantity, and so need not
receive attention here.

Rivers also serve as barriers and boundaries. When a
stream forms the line of division between states their
limits are beyond question; but a river like the Mississippi, which, in its lower reaches, is constantly changing

[1] Deuteronomy xi. 10.

its course, could not be so utilised. In the Republican
period of the Roman power, the Rubicon formed the
line of demarcation between the province of Gallia
Cisalpina and Italia proper. When Cæsar, therefore,
crossed the Rubicon with his legions, it was tantamount
to beginning the campaign ; it was the first act of
war.[1] The Pruth and the Potomac are other rivers
which, in modern times, have separated opposing armies.
A river, from its birth amongst the mountains till its
career is closed upon entering the sea, presents a study
of physical phenomena of a highly interesting character.

Rivers as types of human life. The poet Cowper has compared the passing of time
with the flow of a river ; but, centuries before, Pliny
saw in the rise and development of a river a similitude
to the life of man. "The river springs from the earth ;
but its origin is in heaven. Its beginnings are insignificant, and its infancy frivolous ; it plays among the
flowers of a meadow ; it waters a garden or turns a
little mill. Gathering strength in its youth, it sometimes becomes wild and impetuous. Impatient of the
restraints it meets with in the hollows among the
mountains, it is perhaps restless and turbulent, quick
in its turnings, and unsteady in its course. In its more
advanced age it comes abroad into the world, journey-
ing with more prudence and discretion through cultivated fields ; and no longer headstrong in its course, but
yielding to circumstances, it passes through populous
cities, and all the busy haunts of men, tendering its
services on every side, and becoming the support and
ornament of the country. Now increased by numerous
alliances and advanced in its course, it loves peace and
quiet, and in majestic silence rolls on its mighty
waters until it is laid to rest in the vast abyss."

Sources of rivers. There is something fascinating in the search for the
head-waters of a great river. The origin of the Nile
for centuries evaded discovery, but the search, mean-

[1] To "cross the Rubicon" is from this circumstance chosen to
express the beginning of any important military or other operation.

Fig. 48.—View on the Congo, above Boma.

while unsuccessful, indirectly rewarded the explorers by
revealing the physical geography of the district. It is
not always easy to fix upon the rivulet which develops
into the main stream, and it is not important to do so.

The source. Sometimes an inferior one retains the honour, as in
the case of the Mississippi, the Missouri to the point
of junction being the larger river. More depends, of
course, upon the number and character of the tributaries
than upon the originating stream, a fact not appreciated
by the visitor to the head-waters of the Isis, which
failed during a very dry season, and who imagined that
no water would be found flowing under London Bridge,
and that the vessels in the river would be stranded.
The upper part of a river valley has been named a
combe in Devonshire and the south-west. Having a
supply of water, this would be an eligible position for
a town, and we have from this circumstance Ilfracombe,
Farncombe, Babbicombe, and others. Rivers sometimes flow full-bodied from a spring; but it is remarkable that one so beginning seldom expands into a
first-class river. Here, as in some other cases, small
beginnings with steady growth lead to the greatest
success.

The whole of a district drained by a river and its
tributaries is called its basin, the area of which is a fair
criterion of the importance of a river and its system of
supplementary streams. The Thames basin is estimated
at 5000 square miles; the Rhine at 8000; the Euphrates

Pheno-
mena of
rivers.
and Tigris, 243,000 ; the Nile, over 700,000 ; the Yang-
tse-Kiang, nearly three-quarters of a million of square
miles ; and the Mississippi and the La Plata have each
a basin well over a million of square miles, whilst the
Amazon has nearly two millions. The Congo (Fig. 48),
also in Central Africa, is no doubt a river of the first
class; but its true magnitude has hardly yet been
determined.

The water-
parting.
The line separating two river systems is named the
watershed, or water-parting; in America it is termed

"the divide," as the "Great Divide," the Rocky Mountains. It is not necessary for the water-parting to be high. In Russia, which has a fine hydrographic system, the divide is only a few hundred feet high; but this is sufficient to separate the north-westerly flowing rivers from those having a set to the south-east. In such a case the rivers will be slow, navigable, and useful; and the wild scenery characteristic of the upper course will be omitted.

On the other hand, some rivers take their rise at a great elevation, and, if the mouth be not far distant, they will maintain throughout the impetuous character associated with the upper course, the scenery character-istic of the middle and lower reaches being omitted. A good illustration of this type of river is furnished by Back's River, which rises near the Great Slave Lake in North America. From its source to the Arctic Ocean, into which it flows, the river has the character of a mountain stream, rushing through rocky gorges and forming many waterfalls, finally leaping over a precipice into the sea.

Some rivers flow through defiles. These "narrows" are formed where the waters have cut a passage through solid rock. The Reuss, a Rhine tributary, is of this character, dashing through ravine after ravine, in one of which, three miles long, it descends 1000 feet. An illustration upon a larger scale is afforded by the Columbia, which, rising in the Rocky Mountains, ultimately attains a breadth of several miles. Arriving at a rocky pass, not more than fifty yards across, the whole body of water is forced through at a rapid pace. There is not room for the waters to expand, so that they are of considerable depth; and the surface remains almost unbroken, small boats and even canoes being navigated through them in safety. The rapids of the Shannon, in the county of Limerick, are highly picturesque. The Nile also has rapids of import-ance. In one part of its course there is for a hundred

River scenery.

Narrows and rapids.

miles "a succession of steep descents, and a multitude
of rocky islands, among which the river dashes amid
clouds of foam, and is tossed in perpetual eddies."

Upon the American continent the rapids are commensurate with the size of the rivers. Upon the St.
Lawrence, near its junction with the Ottawa, near
Montreal, for about nine miles of its course, rapids are
formed. These are navigated, but the risk is considerable. It has been the custom, at one point, for the
boatmen to ask the aid of the patron saint of the
district upon their enterprise :—

> "Saint of this green isle, hear our prayers,
> Oh grant us cool heavens and favouring airs.
> Blow, breezes, blow, the stream runs fast,
> The rapids are near, and the daylight's past."

There is a pretty close connection between rapids
and cataracts. In several cases, as upon the Nile,
either name might be applied ; but where the river
bed is broken by a precipice, and the water descends
vertically, a cataract proper is formed. They are
numerous in all the continents, and a mere enumeration
would be beyond the scope of these pages. Niagara is
remarkable for the enormous volume of water which
the river, here three-quarters of a mile wide, pours into
the gulf below with deafening roar, the name being a
corruption of the Indian Oh-nee-gi-ara, which means the
Waterfalls. thunder of waters. The enormous power of these falls
is now being harnessed for useful work by electrical
agency, and it is said that their energy would turn
the machinery employed in the whole of the United
States.

Other falls, as those of Wales and Scotland, are
remarkable for their scenery and surroundings. In
the Vale of Cashmere several of the mountain streams
form cascades of uncommon beauty. In the Lake
The Falls district the Falls of Lodore are familiar through
of Lodore. Southey's description :—

Fig. 49.—The Falls of Clyde.

K

"Delaying and straying, and playing and spraying,
Advancing and prancing, and glancing and dancing:
And so never ending, but always descending,
Sound and motion for ever and ever are blending,
All at once, and all o'er, with a mighty uproar,
And this way does the water come down at Lodore."

The largest English waterfall is on the river Tees, and the scenery of the district greatly enhances the interest. In Scotland the finest example is the Falls of Glomach. Far away from the regular tourist round, in the county of Ross, a mountain stream flows through a wild glen, at one point descending over 100 yards in one unbroken sheet. The Falls of Clyde are also famous (Fig. 49). In Ireland the Falls of Powerscourt, formed by the river Dargle, in the county of

FIG. 50.—Gavarnie Falls

Wicklow, present a fine spectacle when the river is in flood, the curious feature here being the varying rate at which the waters make the descent—some hundred yards—owing to the broken character and altering slope of the rocks over which the waters fall. The Bann, also in the north-east, has a famous salmon leap (Fig. 51).

European waterfalls.   Upon the continent of Europe, cascades and cataracts abound, as in Scandinavia and amongst the Alps. In the Pyrenees are several of great interest, as the Gavarnie Falls, near Mont Perdu (Fig. 50), having the greatest descent in Europe, with scenery of a notable kind. "Let the reader," says Miss Zornlin, "picture to himself an amphitheatre of solid rock, the walls of which rise perpendicularly to the height of more than 1500 feet, terrace above terrace, and surmounted by gigantic columns, 1000 feet in height, also consisting of solid rock, but the capitals of which are formed by the coronets of snow which never quits their summits. In this remarkable and sublimely grand spot rushes down the cataract which forms the source of the Gave de Pau. No other cataract is equal to it, and no other portion of European scenery can be compared to that which forms its birthplace."

The Velino Falls.   In Italy the waterfalls are very fine, as the cascades of Tivoli; but, in certain aspects, the falls upon the Velino River are supreme. Near Terni the stream contracts, and the river, after forming rapids, bounds over a precipice 300 feet deep. The rocks are of marble, and the cataract bears the name of the Cascata del Marmore—the marble cascade. Falling through so great a depth, the waters are accompanied by clouds of spray, which, condensing into rain, gently descend "in an eternal April to the ground."

In Asia are waterfalls surpassing in beauty and majesty any in Europe; but they are seldom visited,

FIG. 52. — Niagara, from Goat Island.

and almost unknown. The river Paber, a tributary of the Ganges, rises, at the height of over two miles, from a large glacier. Many streams unite with it, and at one point in its rapid downward course it falls over a wall of rock to the depth of 1500 feet in two bounds.

The cataract of the Shirawati, in the Himalayas, is in the midst of the richest tropical vegetation; the river, a quarter of a mile wide, and of great depth, forming a fine horseshoe-shaped fall, the total height, with the associated rapids, being over 1000 feet.

American falls.

Upon the American continent the fame of Niagara [1] (Fig. 52) has eclipsed all others, several of which are, however, worthy of mention. Hood's River forms the Wilberforce Falls. Here the waters, with a double leap, bound into an abyss 250 feet below. Rocks rise out of the waters to a great height, the whole forming one of the finest scenes in nature. The falls of the Missouri are second only to Niagara. They extend over twelve miles in continuous succession, and terminate in a cataract of great magnificence. Mexico

South American waterfalls.

and Tierra del Fuego in South America have also numerous interesting waterfalls. Humboldt was enamoured of the cataract of Lequindama, formed by a tributary of the Magdalena. The depth is nearly 200 yards, and an enormous volume of water passes over. A special charm of these falls is the surrounding tropical vegetation, "the luxuriant form of the trees and herbaceous plants, and their distribution into groups, or into scattered thickets; the contrast of the craggy precipices with the freshness of vegetation,

---

[1] The roar of Niagara is heard thirty miles distant; it is computed that 90,000,000 tons of water go over the falls every hour. The upper rock is limestone, but underneath this is a layer of shale, which, less hard, is worn away, causing the limestone rocks to overhang and ultimately to fall. In this way the falls are retreating towards Lake Erie, which may ultimately be emptied, and the geography completely changed.

Fig. 53.—The Murchison Falls, about 120 feet high, from the Victoria Nile or Somerset River, to the level of the Albert Lake.

which stamps a peculiar character on these great scenes of nature."

Of African geography we have yet much to learn. Instead of a waterless waste the interior is found to be a region of mountains and rivers, verdant landscapes, and varied physical charms. Waterfalls are also represented; the Murchison Falls are extremely beautiful (Fig. 53), and South Central Africa contains a cataract which in several of its aspects has no peer. "The most wonderful sight I had witnessed in Africa," says Dr. Livingstone, in his *Missionary Travels*, speaking of the falls of the Zambesi. The first time they were beheld by a European is thus described :—

The Zambesi.

"After twenty minutes' sail from Kalai we came in sight, for the first time, of the columns of vapour, appropriately called 'smoke,' rising at a distance of five or six miles, exactly as when large tracts of grass are burned in Africa. Five columns now arose,[1] and, bending in the direction of the winds, they seemed placed against a low ridge covered with trees; the tops of the columns at this distance appeared to mingle with the clouds. They were white below, and higher up became dark, so as to simulate smoke very closely. The whole scene was extremely beautiful; the banks and islands dotted over the river are adorned with sylvan vegetation of great variety of colour and form. At the period of our visit several trees were spangled over with blossoms. Trees have each their own physiognomy. There, towering over all, stands the great burly baobab, each of whose enormous arms would form the trunk of a large tree, besides groups of graceful palms, which, with their feathery-shaped leaves depicted on the sky, lend their beauty to the scene. As a hieroglyphic they always mean 'far from home,' for one can never get over their foreign air in a picture

[1] The actual height is about one hundred yards. Bright rainbows are seen upon the steam columns. The vapour condenses into rain, which keeps the trees near constantly watered.

of landscape. The silvery mohonono, which in the tropics is in form like the cedar of Lebanon, stands in pleasing contrast with the dark colour of the matsouri, whose cypress form is dotted over at present with its pleasant scarlet fruit. Some trees resemble the great spreading oak; others assume the character of our own elms and chestnuts; but no one can imagine the beauty of the view from anything witnessed in England. It had never been seen before by European eyes, but scenes so lovely must have been gazed upon by angels in their flight. I peered down into a large rent which had been made from bank to bank of the broad Zambesi, and saw that a stream of 1000 yards broad leaped down 100 feet, and then suddenly became compressed into a space of 15 or 20 yards.

"On the left side of the island we have a good view of the mass of water which causes one of the columns of vapour to ascend. As it leaps quite clear of the rock, it forms a thick unbroken fleece all the way to the bottom. Its whiteness gave the idea of snow, a sight I had not seen for many a day. As it broke into (if I may use the term) pieces of water all rushing on in the same direction, each gave off several rays of foam, exactly as bits of steel when burnt in oxygen gas give off rays of sparks. The snow-white sheet seemed like myriads of small comets rushing on in one direction, each of which left behind its nucleus rays of foam." *The Victoria Falls.*

A wonderful cataract.

The river now travels for some miles in a zigzag course between narrow walls of rock, the waters making an indescribable churning noise as they pass into the lower reaches of the river. These falls are the admiration of the natives, who in their poetical language speak of them as "the place of the sounding smoke," and "the place of the rainbows," the spray being generally illuminated with the colours of the spectrum. In the river are several islands thickly wooded with

palms and other tropical trees, one island close to the
brink of the waterfall being employed as a place of
worship. The name Zambesi means *the* river. It is
also called the Leeambye, and what the natives say
of it may be translated—

"Tho Leeambye, nobody knows
Whence it comes and whither it goes."

Familiarity with this cataract does not deaden the
feeling of admiration with which the inhabitants regard
it, and one of the chieftains inquired of the great
missionary-explorer : "Have you smoke that sounds in
your country ?"

Flooding of
rivers.
Other phenomena exhibited by rivers are their inundations, whether annual, seasonal, or irregular. The
flooding of the Nile has been alluded to ; but nearly all
rivers of tropical or subtropical regions are subject to
yearly floods, depending upon the heavy rains which
follow the course of the sun. The Ganges (Fig. 55)
is in flood from June to September, when its carrying power is increased tenfold, and Sir Charles Lyell
states that if a fleet of eighty Indiamen, each freighted
with about 1400 tons' weight of mud, were to sail down
the river every hour of every day and night for four
months continuously they would only transport, from
the higher country to the sea, a mass of solid matter
equal to that borne down by the Ganges during the
flood season. The Yellow River, known as "China's
Sorrow," occasionally floods its basin (Fig. 54). The
Tigris exhibits a semi-annual flood, first in April, on
account of the melting of the snows of the Armenian
mountains, and again when swollen by the autumn
rains. The Mississippi has heavy floods in the spring;
and as in many parts of the lower course the surface is
upon a level with the surrounding land, serious dam-
age sometimes occurs. The Indus, the great western
river of India, to which it imparts its name, has occasionally, as in the year 1841, floods of a tremendous

character. Nearer home the French rivers not infrequently give trouble. In the year 1818 the Rhone
valley was the scene of a devastating flood, by which
rocks were removed, forests, houses, and bridges swept
away, landmarks obliterated, and the plain of Martigny
strewn with wreckage. Rivers are now, however, under
some control, and the method of "training" rivers is
much practised.

FIG. 54.—"China's Sorrow": the Yellow River in flood.

The effect of cultivation is to increase the liability
to flooding where rivers constitute the main drainage
of a country. Storm water more quickly reaches the
river as marshes, which act as flood moderators, are
removed. Where a river flows through a lake, the
river below the lake has a more constant flow, a large
lake absorbing storm water without the level being
much raised.

It has been mentioned that the principal rivers of

River
mouths.

England are united by canals. There are, however,
several examples of neighbouring river systems being
connected by tributaries of the rivers themselves. If
taking place in the lower course, such a union is of
little commercial importance; but where navigable
rivers are united in their upper reaches, a useful water-

Bifurca-
tion.

way is formed. The Tiber and the Arno are thus
joined, and also the Mahanuddy and Godavery; but
the finest illustration is where the upper waters of the

FIG. 55.—The Ganges at Derali.

Orinoco and the Amazon are connected by the river
Cassiquiare, some thousands of miles inland.

British
rivers have
estuaries.

Rivers enter the sea either by a single trumpet-
shaped opening, or by a number of mouths. The
Thames, the Severn, the Mersey, and other rivers of the
British Isles are of the former class, which is to the
advantage of our commerce, as their openings are wide
and unobstructed. Some have a tendency to fill up
their mouths, to become deltoid, as the Ribble, the
Great Ouse, and the Mersey, which require dredging.

Of rivers forming deltas there are many examples.
The Volga has seventy mouths; the Indus enters the

sea by several muddy openings, and the Rhine delta
includes a large part of the Netherlands. The Nile
delta begins 90 miles from the sea and extends 200
miles along the coast, the two principal mouths being
the Rosetta and the Damietta, and it has been computed
that this great deltoid tract has been forming for Deltoid
13,000 years.   The Mississippi has a delta 12,000 rivers.

Fig. 56.—On the Amazon : Tropical Vegetation.

miles in area, and the Orinoco has an important one,
as is also the case with the Zambesi ; but perhaps the
finest instance of a delta is that of the Ganges and
the Brahmapootra, which covers 22,000 square miles,
and is also of great thickness.   Deltoid land rising
but slightly above the sea is liable to inundation, the
floods at the Ganges mouth being occasionally very
destructive.

Besides the rivers already referred to there are Giant
others of gigantic proportions, or of unusual character- rivers.

The
Amazon.

istics. The Amazon (Fig. 56) has the largest basin
—over 2,000,000 square miles. The river varies
from one mile to two miles wide in its upper course,
and develops into a sea-like stream, being 100 miles
across before reaching the Atlantic, into which it
pours a volume of water tenfold greater than any other
river. The shade temperature often reaches 100°
Fahr., and, with a rainfall of 200 inches in some parts,
vegetation is prolific, a mighty forest occupying the
basin. There are no less than twenty important tributaries, and it is computed that the system of the
Amazon has 50,000 miles of navigable water. The tide
ascends for 400 miles with a bore the waves of which
are from 10 to 15 feet high. The dimensions of this
river are easily written, but it is hardly possible for the
untravelled Englishman to realise them.

The Po, which drains the plain of Lombardy, a
region as large as the whole of England without Wales,
flows in certain parts of its course at a higher level
than the towns and villages near, the waters being
banked out like the sea in Holland. The Danube has
a course of 1700 miles, with sixty navigable tributaries, and a basin of over 250,000 square miles. The
La Plata of South America has a basin of over
1,000,000 miles, and enters the ocean by an estuary
200 miles wide, its waters being perceived more than
100 miles at sea.

The Mississippi, with the Missouri and its tributaries, forms a navigable water-course more than
sufficient to engirdle the earth at the equator. The
course of the main stream is extremely circuitous.
It flows through a region of remarkable fertility,
the basin of this river constituting nature's greatest
gift to the United States. The basin is over 1,500,000
square miles, and it is said that one may steam up
this river for 2000 miles without its width sensibly

The St.
Lawrence.

diminishing. The St. Lawrence, the great river of
the north-east, has numerous islands, and an estuary

FIG. 57.—Windsor Castle from the Thames.

100 miles wide ; and forms a waterway which is one
of the greatest natural advantages of the American
continent.

Of English rivers, the Thames, from any point of
view, occupies the premier place (Fig. 57). In magnitude it is a mere rivulet compared with the giant
streams of the great land masses ; but in some respects
it deserves to rank with the first.   Its source in
the Cotswold Hills is only about 100 yards above the
sea.   Its current is consequently slow, and it is navigable up stream for nearly 100 miles. The tide, which
keeps the channel open, is felt for over fifty miles,
reversing the current twice daily, and greatly aiding
navigation. Its tributaries join from both banks, so
that the principal southern counties are connected by
its waters, and it enters the sea by a wide but somewhat obstructed estuary. The Thames, in fact, is the
dominating physical feature of the south - eastern
counties, and in the "making of England" it has
played a prominent part. Its scenery also is not surpassed by any British river (Fig. 58), and its associations are a precious heritage to the English people.
The prospect, for example, from Richmond Hill along
the valley, with its "eyots, skiffs, and swans"; with
the beech-clad hills of Buckinghamshire and majestic
Windsor on the far horizon, is, on a clear day, one of
the great sights of English landscape, which Thomson's
verse and Turner's pencil have made known.   There is
not space here to recapitulate the historic scenes which
this river brings to mind; but there is one place of
commanding interest.   It is the scene of the conference
on that memorable morning of June 1215, when "the
army of God," with Robert Fitzwalter, the flower of
the English nobility, and Stephen Langton, Archbishop
of Canterbury—but "more Englishman than priest"—
met King John with his retinue, and obtained the
reluctant monarch's signature to the charter of English
freedom. The army of the barons, marching from

FIG. 58.—On the Thames : Nuncham Courtney.

L

Staines, met the king in the green meadows by the Thames. There is no monument to keep Magna Charta and Runnymede in remembrance. All that can be seen are the cattle quietly grazing, and the water-plants gently moving with the flow of the river.

Fig. 59.—The Black Rock of Novar, Cromarty Firth.   A chasm eroded by a stream in Old Red Conglomerate.

VI

LAKES AND THEIR LESSONS

" By Killarney's lakes and fells,
    Emerald isles and winding bays,
Mossy banks and woodland dells,
    Memory ever fondly strays.
Bounteous nature loves all lands,
    Beauty wanders everywhere,
Footprints leaves on many strands,
    But her home is surely there !
        Angels fold their wings and rest
        In that Eden of the west.
        Beauty's home, Killarney !
        Heaven's reflex, Killarney ! "

THE divisions of the land have their counterpart in
the waters. The continental masses have for their
opposites the ocean expanses ; the peninsulas and
capes, the bays and gulfs ; and the isthmuses connecting the land areas, the straits which unite the seas ;
whilst islands, which break up the monotony of the
great waters, have their analogue, as it were, in the
lakes which, like mirrors, brighten up and burnish the
land.

The extent of lake surface is estimated at about a
million of square miles ; but the importance in the
world's economy of these "islands of the land" is more
than commensurate with their magnitude.

Lakes greatly enhance the beauty of scenery, some

Uses of
lakes.

Their
scenery.

of the more favourite holiday and health resorts being in their vicinity, as the lake districts of England, Scotland, and Ireland. Moore has sung of the charms of the Irish lakes, and Sir Walter Scott of the beauties of those of his native land, as thus of Loch Katrine (Fig. 60) :—

> "One burnished sheet of living gold,
> Loch Katrine lay beneath him roll'd ;
> In all her length far winding lay
> With promontory, creek, and bay,
> And islands that, empurpled bright,
> Floated amid the livelier light."

The Swiss and Italian lakes are also admired ; and, in a smaller measure, the meres and pools scattered through the English counties greatly enhance the scenery, as in the extensive plain which forms a considerable part of the county of Cheshire.

Other uses
of lakes.

Lakes assist in forming commercial waterways, especially when arranged as a chain. Thus the Bitter Lakes, which are really inland seas, aided the construction of the Suez Canal, navigation through them being directed by buoys ; and the dull, marshy lake of Menzaleh at the Mediterranean end similarly assisting, the canal is now animated with the ships of an important commerce. Nearly 70 miles of the whole length of 100 miles are through lakes.

Promote
internal
communi-
cation.

Similarly, the succession of long lochs stretching across the Great Caledonian Glen from Loch Linnhe to the Moray Firth form a large part of the Caledonian Canal. Loch Ness in this chain is very deep, and consequently in the severest weather is not frozen over.

This kind of service is also well shown in North America, the northern portion of which is a veritable "land of lakes." From the Great Slave Lake, through Athabasca Lake, Deer Lake, and Winnipeg, there is, by their means, water communication on to the Lake

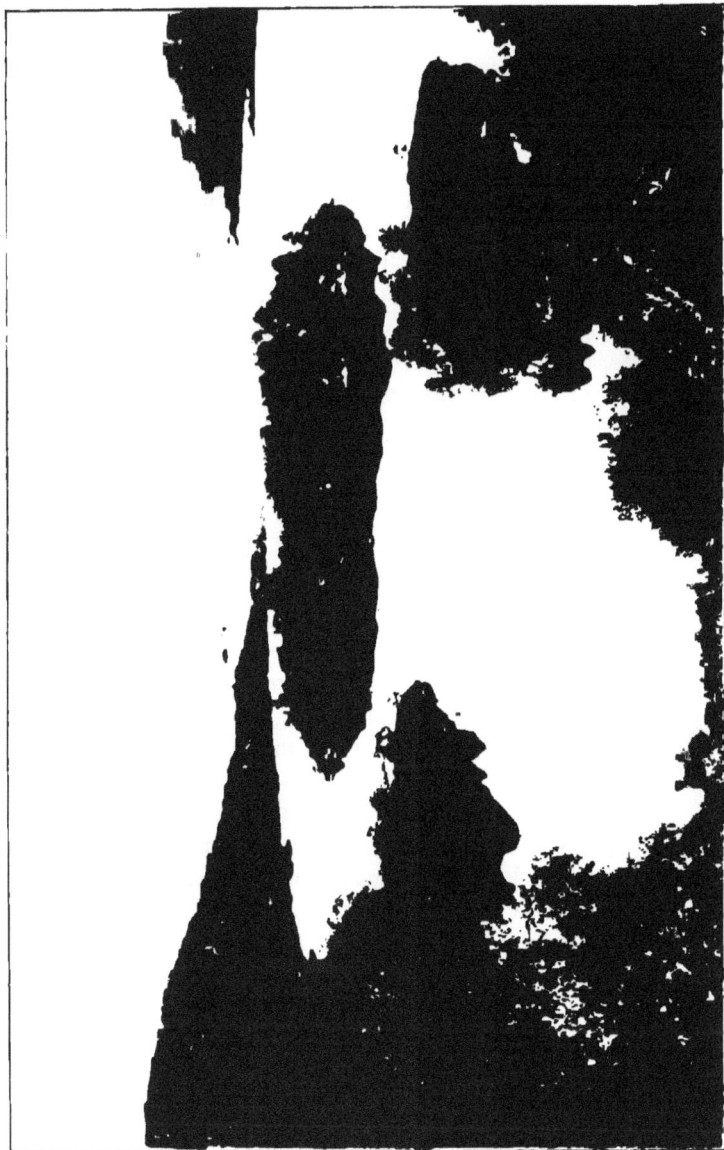

Fig. 60.—Loch Katrine and Ellen's Isle.

of the Woods, and extending to the splendid system [1]
—Superior, Michigan, Huron, Erie, and Ontario, to
the St. Lawrence—by which waterway the New and
the Old World conveniently exchange their products.
Then some single lakes are extensive enough to afford
water carriage between places that would communicate
less easily by land. The remarkable rise of Chicago
(pronounced Shikargo) is largely owing to its position
upon the verge of Lake Michigan, and upon the line of
water and railway stretching across the country. The
rapid growth of this city, "the Venice of America," is
indicated by the fact that it is not marked upon maps
issued half a century ago.

Improve climate. Lakes favourably influence climate and season.
Their waters maintain a more equable temperature
than the land. In tropical regions the lake breezes
temper the heat of day and reduce the cold of night,
which in these parts is never short. In the temperate
zones they also moderate the extremes of temperature
of countries possessing a continental climate,[2] especially
when deep as well as large.

Lakes further act as "flood moderators" (as described
in speaking of rivers), receiving storm water without
rapidly rising in consequence, and, owing to this, great
rivers which flow through lakes maintain an even
current, and are preserved from such devastating floods
as occur, for example, upon the great Chinese rivers
which have not this absorbing power.

Induce variety of animal and vegetable life. Lakes are the means of developing a special flora
and fauna. Aquatic plants spring up, affording cover
for water-loving animals, which impart life and beauty
to regions that would otherwise be dreary and desolate.

Yield stores of salt. Other lakes deposit beds of salt, often over a large

[1] These are of the nature of freshwater seas—Superior, 32,000
square miles ; Michigan, 24,000 ; Huron, 20,000 ; Erie, 9000 ; and
Ontario, 6000. Total, 94,000 square miles, or a larger area than the
whole of the British Isles.

[2] Continental climates are exemplified by Russia and Canada,
which have a continuously hot summer and a long severe winter.

Fig. 61.—Loch Leven.

area, and at a more rapid rate than might be supposed ; and where the waters are diminishing through excessive evaporation, or the feeders become enfeebled from any cause, their floors will be ultimately revealed, and their valuable stores made more available for man's use.

Scotland is a land of lakes (Fig. 61). The areas known as "the lochlands" are the sites of former lakes ; and their beds, now dried, have an interesting story to tell to the antiquarian and the geologist ; for the floors of lakes, unlike the floors of the dwellings of man, thicken with age (the various layers follow in stratigraphical order, the nearer the surface the more *The lesson of the lochlands.* recent in time) and bear record of distinct periods in the earth's history. Objects dropped in the waters of the lakes become imbedded ; and when, it may be after long ages, the lake dries up, its bed, when dug into, may prove to be a complete museum of relics, which, brought under the criticism of the expert, will afford evidence intelligible and incontrovertible of the life of times that would otherwise be involved in obscurity.

Extraordinary testimony of this kind has been given by certain lakes, whose waters have been reduced by evaporation or partly drained off by agricultural opera-tions. The ruins of ancient dwellings and even whole villages have been revealed in several of the Swiss *Teaching of the Swiss lakes.* lakes. These dwellings of pre-historic people are cir-cular in plan, and formed somewhat after the pattern of a beehive. The village was connected with the shore by a narrow causeway. At the bottom of the lake have been found the remains of these early people. The Norman baron surrounded his castle with a moat as a defence against his foes ; but the " Lakers " planted their houses upon piles in the midst of the water, hoping thus to escape the attacks of wild beasts, with which they had to contend, and of their human enemies, who were hardly less savage. In the year 1856, from Lake Morseedorf, six miles from Berne, and at different times from other lakes of the district,

very complete evidence has been obtained of the char- *The Lakers.*
acter of these people and of the life they led. Their
implements, generally of serpentine, are small but
tastefully carved and accurately shaped. Bones and
teeth, and the remains of the domestic animals
with which they were associated, as the ox, hog,
goat, cat, and dog, and of the wild creatures to which *A pre-
they gave chase, as the urus, bear, wild boar, fox, historic
and beaver, have been collected, and are preserved people.*
in the Swiss museums. These sub-lacustrine finds also
include pottery (rudely fashioned by hand before the
introduction of the turning-wheel), daggers, bone
needles, saws, arrow - heads, and even specimens of
clothing. Fragments of the food they ate—generally
seeds and fruits—have also been fished up; and, by
careful study of these relics, an insight has been gained
as to their manners and customs and modes of sepul-
ture. Quite a history, in fact, could be written of these
"Lakers," who are computed to have lived in the
newer stone-using period (Neolithic). The Swiss lake
explorers have hinted that, should the waters of our
English lakes at any time be drawn very low, they may
afford similar evidence of the life of man in the earliest
times.

Freshwater lakes often abound in fish. The in- *The harvest
habitants near them reap a harvest of the waters the of the
seed for which they have not sown, for fish multiply lakes.*
independently of human agency. In the New Testament we read how the Sea of Galilee was surrounded
with little fishing towns, and how the disciples were
chiefly fishermen; and the newly - discovered freshwater lakes of Equatorial Africa abound in fish, which
afford sustenance to the native races around them.

To the student of nature the character and mode of *Classifica-
formation of lakes opens up an interesting chapter in tion of
physiography. Lakes have had a beginning, and a lakes.*
growth, and will have an end. By careful observation
much may be learnt as to the process, and such a habit.

of inquiry will give zest to travel whether at home or abroad.

A preliminary step is to classify. An obvious arrangement is into salt and fresh water lakes. The former, whose waters are sometimes more saline than the ocean, are, in many cases, the relics of a vanished ocean. When the sea bed was raised to the dignity of dry land, the deeper parts, although lifted up, would still hold water; and if sufficiently fed by rivers and springs, and not exhausted by evaporation, or filled up by deposits from affluents entering it, would remain to this day. Good examples of this kind of lake are the Caspian and the Sea of Aral.

On the other hand are freshwater lakes like those of North America. But nature refuses to be bound by arbitrary classification of man's devising, and there is an intermediate group whose waters are neither salt nor fresh, but only brackish.

A fourfold grouping of lakes, independent of the character of their waters, has been made: (1) Lakes having neither surface feeders nor outlets; (2) Lakes possessing outlets but no feeders, these being replenished by springs in their beds which supply the loss by evaporation; (3) Lakes having feeders but no outlets, the balance being maintained by evaporation where the lake is not increasing; and (4) The more numerous class of lakes which have both affluents and outlets, which are at once lakes of transmission and of reception.

Formation of lakes.   Another view is to consider lakes with reference to their geological surroundings. When occurring in mountainous regions they are characterised by wild and romantic scenery, bare rocks forming their banks, and rocky islets rising from their waters. They are generally small, deep, and clear, and illuminate like "gems amongst the mountains" the often sombre scenes in which they are placed. In contrast with mountain lakes are those of plains and of alluvial or secondary formations. They are generally uninteresting as to

scenery, their shores being low and tame. Such lakes
vary much in size, being swollen by the rains, or
reduced by long-continued drought, and they not infrequently give off unwholesome exhalations.   Such
lakes are found along the south Baltic coast, and around
the Caspian Sea.

The mode of formation of lakes has been various.
It has been shown that some of them are the remains
of a departed ocean ; others are due to the falling
in of cavern roofs ; some have been made by stone
avalanches forming a dam across the river ; glaciers
have also worn out basins in the rocks, which have
remained filled with water when the glaciers themselves
have passed away ; and, less frequently, a lava stream
has intercepted the flow of a river and formed a
lake.   Where the valley through which a river passes
widens, a long lake is formed, many having this
character and mode of origin.   Lakes amongst the
mountains are often due to the action of glaciers
unequally wearing the rocks over which they ground
their way in their downward path towards the sea,
the moraine matter they carried forming a weir or
bank, and thus arresting the water flowing from the
dissolving glacier.

The searching out the precise mode of formation of
a lake may, as has been hinted, give pleasant occupation
to the scientific tourist, and, as nature's laws are universal, much may be learnt without travelling far.
The spirit in which such a research may be undertaken has been described by Kingsley. Speaking of   *Origin of*
an angler at one of the Welsh lakes waiting for   *a Welsh*
a "fishing curl," it is suggested that it would have   *lake.*
been better, "instead of falling asleep, to have worked
quietly round the lake side, and asked of nature the
question, How did this lake come here ? what
does it mean ? It is a hole in the earth.   True, but
how was the hole made ? There must have been
huge forces at work to form such a chasm.   Probably

the mountain was actually opened from within by an earthquake. . . . Yet, after all, I hardly think the lake was formed in this way, and suspect that it may have been dry for ages after it emerged from the primeval waves and Snowdonia was a palm-fringed island in a tropic sea. You see the lake is nearly circular; on the side where we stand the pebbly beach is not six feet above the water, and slopes away steeply into the valley behind us, while before us it shelves gradually into the lake ; forty yards out, as you know, there is not ten feet of water ; and then a steep bank, the edge whereof we and the big trout know well, sinks suddenly to unknown depths. On the opposite side that vast flat-topped wall of rock towers up shoreless into the sky, 700 feet perpendicular ; the deepest water of all we know is at the very foot. Right and left two shoulders slope down into the lake. Now turn round and look down the gorge. Remark that this pebble bank on which we stand reaches some fifty yards downwards; you see the loose stones peeping out everywhere. We may fairly suppose that we stand on a dam of loose stones 100 feet deep." The pebbles are found upon examination not to be of the rock of the country, which is Snowdon slate, but of syenite, an igneous rock, "once upon a time in the condition of hasty-pudding, heated to some 800 degrees of Fahrenheit." The conveyance of these stones and the manner in which the lake was formed is thus told : "How did those pebbles and boulders get 300 yards across the lake ? hundreds of tons, some of them three feet long ; who carried them across ? The old Cymry were not likely to amuse themselves by making such a breakwater up here in No-man's land, 2000 feet above the sea ; but somebody or something must have carried them ; for stones do not fly, nor swim either." After discussing several theories it is seen that there is only one natural agent which could have

VI LAKES AND THEIR LESSONS 157

hollowed out the lake basin and carried and placed the
bank of pebbles in position. " Yes ; ice ; Hrymir the
frost-giant, and no one else. Over the face of this cliff
a glacier has crawled down from that névé, polishing
the face of the rock in its descent; but the snow,
having no large and deep outlet, has not slid down in
a sufficient stream to reach the vale below, and form a
glacier of the first order, and has therefore stopped
short on the other side of the lake, as a glacier of the
second order, which ends in an ice cliff hanging high
upon the mountain-side, and kept from further progress
by daily melting. . . . The stones which the glacier
rubbed off the cliff beneath it, it carried forward, slowly
but surely, till they saw the light again in the face of
the ice-cliff, and dropped out of it under the melting of
the summer sun, to form a huge dam across the ravine ;
and the Ice Age past, a more genial climate succeeded,
and névé and glacier melted away ; but the ' moraine '
of stones did not, and remains to this day the dam
which keeps up the waters of the lake."

The changes going on continually amongst the English lakes Miss Martineau has described :—

" The margins of the lakes never remain the same
for half a century together. The streams bring down
soft soil incessantly ; and this more effectively alters
the currents than the slides of stones precipitated from
the heights by an occasional storm. By this deposit
of soil new promontories are formed, and the margin
contracts till many a reach of waters is converted into
land, inviting tillage. The greenest levels of the smaller
valleys may be seen to have been once lakes ; and no
one who looks down upon Grasmere can have any
doubt as to what was once the extent of the waters.
And, while nature is thus closing up in one direction,
she is opening in another. In some low-lying spot a
tree falls, which acts as a dam when the next rains
come. The detained waters sink, and penetrate, and
loosen the roots of other trees ; and the moisture which

How a
lake
changes.

they formerly absorbed goes to swell the accumulation, till the place becomes a swamp. The drowned vegetation decays and sinks, till the place becomes a pool, in whose margin the snipe arrives to rock on the bulrush, and the heron wades amongst the water-lilies to feed on the fish which come there, nobody knows how. As the waters spread they encounter natural dams, behind which they grow and deepen, till we have a storm among the hills, which tempts the shepherd to build his hut near their brink. Then the wild swans see the glittering expanse in their flight, and drop down into it, and the water-fowl make their nests among the reeds. This brings the sportsmen, and a path is trodden on the hills, and the spot becomes a place of human resort."

The English lakes.   The Lake district has scenery of great attractiveness. The combination of rocks and woods with the waters is one which does not easily pall upon the visitor. The naturalist here finds a rich and varied flora, and the geologist has abundant scope for his hammer. Windermere (Fig. 62) is the finest English lake, affording a drive of some thirty miles around its shores ; its waters are deep and clear. The lake has two important feeders, and its superabundant waters are discharged into the Solway. Heavy rains make but little impression upon this capacious lake. The landscape is soft and pleasing in outline, but is wanting in the boldness which marks some of the smaller of the English lakes. The best part is near Bowness, near the centre, where is the finely-wooded island of Belle Vue. Grasmere and its valley are of uncommon beauty, all the details of which have been faithfully described by Wordsworth. Rydal Water (Fig. 63), Ulleswater, Elterwater, Coniston Water, and Derwentwater, other lakes of the district, are all well worth visiting for their scenery, and as exemplifying important geological changes. Of Coniston Water it has been said that "nowhere else, perhaps, is the grouping of mountain peaks and the indentures of their

Windermere.

Fig. 62.—Windermere.

Coniston
Water.

recesses so striking; and as to the foreground, with its glittering waterfalls, its green undulations, its distinguished woods, its clear lake, it conveys the strong impression of joyful charms, of fertility, prosperity, and comfort, nestling in a bosom of the rarest beauty." The impression of magnitude when viewing the Swiss lakes and their mountain surroundings is diminished by the unusual transparency of the air, distances appearing less than they really are, whilst the scenery of the English lakes gains in effect through this want of perfect clearness in the air. The English lakes have a further advantage in that their surfaces are often perfectly calm, and the way they mirror the scenery is perfect.

. A paper read before the Royal Geographical Society, Midsummer 1894, describes the lake district of North-western England as a definite geographical unit—a circular mass of elevated land, highest in the centre, and surrounded by long narrow valleys radiating from the high centre like the spokes of a wheel. The lakes are generally long and narrow, and are either of the shallow type, as Derwentwater and Bassenthwaite, averaging 18 feet in depth, and shallowed by glacial accumulations, or deep,—40 feet or more—as Ulleswater and Windermere, the bottom of several being below the level of the sea.

Scottish
lakes.

Scotland is a land of picturesque lakes. Loch Lomond is the largest, containing 45 square miles. Its scenery is improved by islands, whilst the shores are thickly wooded, and the heights of Ben Voirlich and Ben Lomond form an imposing background to the view.

Killarney.

The Irish lakes in the north-east are unattractive; but, in the south-west, those of Killarney are the admiration of visitors, and the poet is naturally eloquent in their praise.

The
Caspian.

Of salt lakes the Caspian receives the designation of sea, containing over 130,000 square miles, an area considerably larger than that of the whole of the British Isles. The Volga, which is larger than any other

FIG. 63.—Rydal Water.

M

European river, and several others, as the Ural and the Kur (the ancient Cyrus), here discharge their waters; but evaporation from the large surface is in excess of the supply of water, and this inland sea is steadily diminishing, so that ultimately its bed may be laid bare, and the enormous deposits of salt it is forming be rendered more available. This must be a long process, as in some parts the Caspian is half a mile deep. Other salt lakes are the Aral, in the same region; the lakes in the south-east of European Russia, Lake Eltonsk being the saltest of the group, having a salinity

The Dead Sea. of 39 per cent; and the Dead Sea, which, although small, (39 miles by 9), is interesting historically as well as for its physical character. The waters are seven-fold salter than the sea, and are also bitter, pungent, and nauseous. The landscape is dreary in the extreme. Vegetation is deficient, and no animal life is found either in the waters or around the shores, whilst sombre mountains form its boundary. " The Sea of Lot " this lake has been named, as occupying the site of the wicked cities of the plain; and for this and other reasons its waters have long had an evil reputation. Around these shores, according to Josephus, grew the apples of Sodom, which were said to turn to ashes on the lip; and which are referred to by Milton "like those which grew near that bituminous lake where Sodom flamed." Modern travellers confirm the estimate of the ancient writers as to the appearance and character of this lake, which is in marked contrast with the amenity which lakes in general give to a district. Other lakes of this class, but whose waters are less saline, are the Great Salt Lake of Western North America, and Uros in Bolivia, which has an area of 2000 square miles.

Fresh-water lakes.

The Holy Lake. Amongst freshwater lakes, besides those of North America, previously referred to, may be mentioned Lake Baikal, "the Holy Lake" of Siberia, on the northern slope of Asia, and 1800 feet above the sea. It is 400 miles long by 20 to 50 broad, and with a

FIG. 64.—View up the Valais from the Lake of Geneva.

surface of 14,000 square miles ; but the details of its geography are not yet perfectly known. Numerous rivers discharge their waters into it, and the principal outlet is by the river Angara into the Yenesei and onwards to the Arctic Ocean. Around this great lake the igneous forces of the earth are in great activity.

Lake Geneva.

Of European lakes Geneva is first (Fig. 64) for scenery if not for size. It has an area not far short of 100 square miles, and is elevated a quarter of a mile above the Mediterranean. Its waters are blue and deep. In form it is a crescent. Several rivers enter it, and the Rhone discharges its superabounding waters. This lake has two special characteristics. It is occasionally influenced by a sub-aqueous wind, which breaks through the waters, producing a commotion that is dangerous to small boats ; and to an abrupt rising of the level, sometimes to the height of several feet, without visible cause.

Lake Constance.

Closely following Lake Geneva in size and importance amongst the Swiss lakes is Lake Constance. Like Geneva it is in some parts very deep, nearly half a mile, and it also exhibits the phenomenon known as the "Seiches," or the sudden rising and falling of the waters—on one occasion to the height of two or three feet four or five times in an hour. Over fifty streams enter this fine lake, and the Rhine discharges the overflow waters.

Strange lakes.

Lakes yielding borax.

The borax lakes of Tuscany are also very curious. They occupy an area of about thirty square miles near one of the volcanic centres of Italy. Boiling water and vapour, accompanied by a sulphurous smell, issue from them, causing the air to become hot. These lakes were long held in fear by the peasantry, who spoke of the region as the mouth of the infernal regions ; but modern science has tamed their powers, and their waters have yielded tons of boracic acid for manufacturing purposes. In Central Asia, as in Thibet, are other lakes yielding borax in large quantities. In

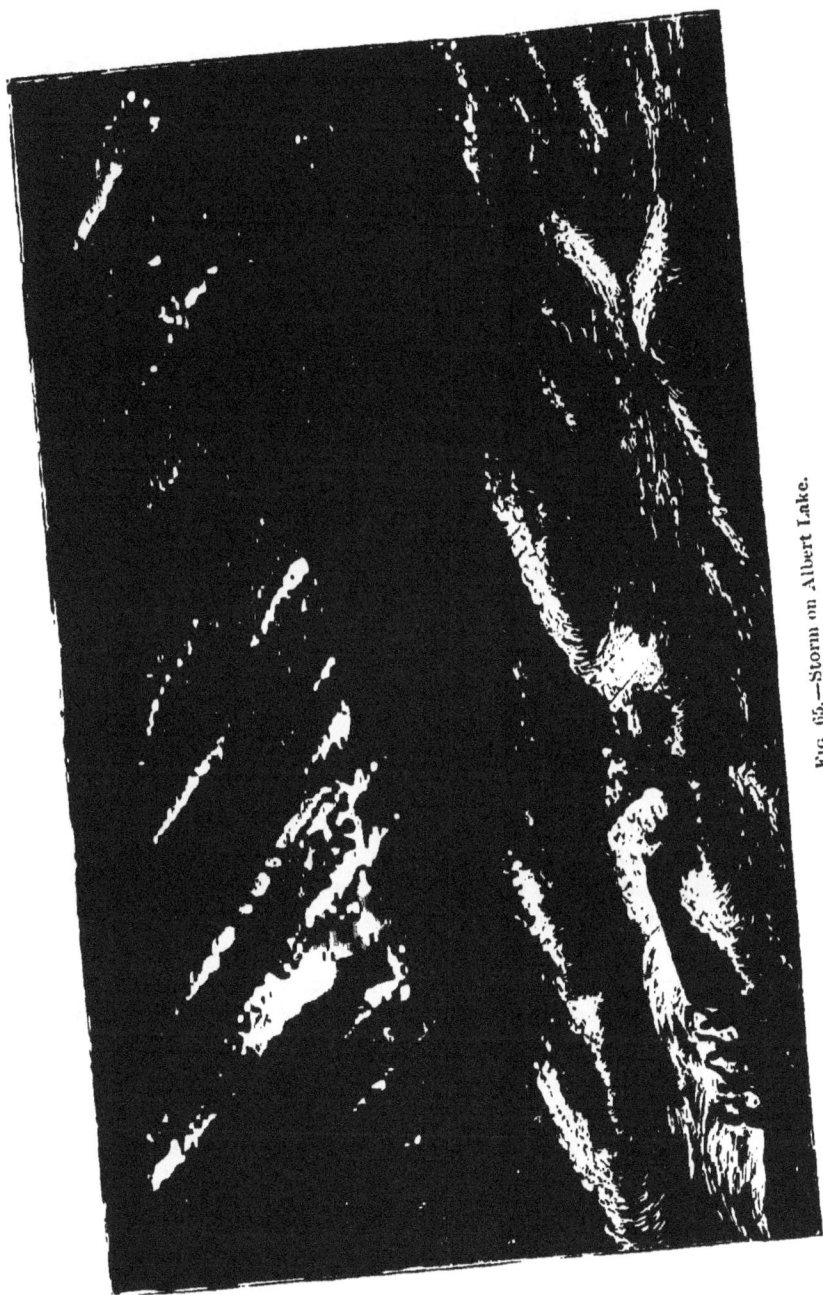

Fig. 65.—Storm on Albert Lake.

Egypt are lakes producing natron or carbonate of soda in considerable quantity. This material is spoken of under the name of " nitre " in the Bible by the prophet Jeremiah, and also in the Proverbs: " As vinegar upon nitre, so is he that singeth songs to a heavy heart."

African lakes.   Africa is now known to possess lakes of the utmost importance. Besides those long known in Egypt, and Tchad, on the southern edge of the Sahara, the travels of Dr. Livingstone, of Stanley, and other explorers have revealed the existence of great bodies of fresh water, in the equatorial regions, of extreme value in the future development of " The Dark Continent." From the Albert Nyanza (Fig. 65), which is at least 300 miles long, and which lies directly under the equator, the river Nile takes its rise, and thus a mystery of African geography has been solved. From the Victoria Nyanza, which is not quite so long, but more compact in form, another branch, the White Nile, is thought to originate. The Albert Nyanza is surrounded by mountains, some of which rise to the height of two miles, and the whole region is one of great diversity, altogether unlike the conception of half a century ago, when this part of Africa was marked upon maps as

Tanganyika.   unexplored country. Another of the vast lakes of Equatorial Africa is Tanganyika. It is fresh water and abounds in fish, and around the shores are the fishing villages of the natives, and places such as Ujiji, of considerable importance in this part of the African continent. Numerous rivers contribute to these waters, and one of the outlets, the Lakuga River, flows into that immense river the Congo. Several other headwaters of this magnificent stream arise from other lakes near. The Congo in its course occasionally widens out, forming lakes hundreds of miles in length. Numerous falls occur along this river and the lakes to which it gives rise, but by the construction of canals such obstacles to navigation may be avoided, and the hydrographic system, which is only just beginning

FIG. 66.—The Pitch Lake.

to be understood, should prove of the first importance in the civilisation and development of Central Africa. The difficulties of exploration and the obstacles to be encountered in traversing this waterway of Africa are described in Mr. Stanley's journal: "From a list of seventy-four falls, cataracts, and rapids, which we had to encounter in our long descent of the Livingstone to the western ocean, it may well be imagined that some were of such a nature that they required great study to discover the means to pass them, while others, again, compelled us to adopt the only plan left—that of undergoing the enormous labour of hauling the canoes up the mountains. One of the latter was the Inkisi Falls. Inviting my friendly natives to my aid, we buckled on the two largest canoes, while the weaker Waugnana cut a road through the forest that covered the slope. At this particular part the mountains rose in a terrace with a steep face, 300 feet high, and then ascended, by a more gradual slope, 1200 feet to the summit of the tableland. Before we could be said to pass the fall we had to drag our canoes up to the summit of the tableland, 1500 feet above the river, then over the tableland, a distance of three miles, and down again to the river, with a descent of 1500 feet."

Of lakes exhibiting special characteristics may be named Titicaca in Peru, which is situated at an elevation of two miles above the sea-level. Lake Van in Armenia is also over one mile high; but the loftiest mountain range, the Himalayas, is almost without lakes, the largest having only the dimensions of Loch Lomond. A very curious one is the pitch lake of Trinidad (Fig. 66).

Several lakes exhibit the curious phenomena of floating islands, as in Sweden, Germany, and Italy. Of one of these islands, in Lago di Bassanello, in the last-named country, Pliny has given a description: "Several floating islands swim about on it covered with reeds and rushes, together with other plants, which the

(margin notes)
Difficulties of exploration.

Curious lakes.

neighbouring marsh and the borders of the lake produce. These islands differ in their size and shape;
but the edges of all of them are worn away by their
frequent collision against the shore and each other.
They have all of them the same height and motion, and
their respective roots, which are formed like the keel
of a boat, may be seen hanging down in the water, on
whichever side you stand. Sometimes they move and
cluster and seem to form one entire little continent;
sometimes they are dispersed in different quarters by
the winds; at other times, when it is calm, they float up
and down separately. You may frequently see one of Floating
the larger islands sailing along with a lesser joined to islands.
it, like a ship with its long-boat, or perhaps seeming
to strive which shall out-swim the other. The sheep
which graze upon the borders of this lake frequently go
upon these islands to feed without perceiving that they
have left the shore, till they are alarmed by finding
themselves surrounded with water; and in the same
manner, when the wind drives them back again, they
return without being sensible that they have landed.
I have given you this account because I imagined it
would not be less new nor agreeable to you than it was
to me, as I know you to take the same pleasure as
myself in contemplating the works of nature." Lake
Gerdau in Prussia has, upon a floating island, pasturage for a hundred cattle, and another lake in the
same country has trees of the largest size established
upon one of the same formation. At the mouth of the
Mississippi the trees brought down in flood-time have
made an island over ten miles long.

The waters of lakes exhibit diverse shades. The Colour of
Great Bear Lake is bright blue, and is of great trans- lakes.
parency in some parts, a small white object being seen
at the depth of nearly 100 feet. The waters of
Lakes Superior and Huron are also uncommonly clear,
and in some of the Scandinavian lakes the water appears
nearly as clear as the air, so that it is sometimes

difficult to distinguish between an object and its reflection when the surface is calm.   A farthing, it is said, has been seen in Lake Wetter in Sweden at the depth of over 100 feet.

Temporary lakes.   Several lakes exhibit a curious periodicity, as Lake Zirknitz in Illyria.   It is sometimes full of water

FIG. 67.—Mirror Lake, Yosemite Valley.[1]

for several years in succession, and then the waters entirely disappear.   A small lake in Greece behaves similarly, the waters occasionally forsaking it, so that the bed may be cultivated.

On the whole, it is thought that the lakes of the world have decreased in size during the historic period,

---

[1] In this view several objects are seen through the water alone and mix with the reflection.

Fig. 68.—Reflected Scenery.

and with the lapse of time their beds may be added
to the dry land. Several lakes give evidence of once
having occupied a large area, the Great Salt Lake of
America, for example, showing several old shore lines
at which the waters formerly stood.

Reflected scenery.   The reflective power of the surface of lakes, just
referred to, is best seen in the Mirror Lake of the
Yosemite valley (Fig. 67). It is only by the reflection that the observer perceives where the line of the

FIG. 69.—Terraces of Great Salt Lake, along the flanks of the Wahsatch
Mountains, south of Salt Lake City, indicating the shrinkage of the waters.

surface really is. Coleridge's analysis of the pleasure
derived from reflected scenery amongst the English lakes
would be appreciated by the traveller : " The imagination by this aid is carried into recesses of feeling otherwise impenetrable. The heavens are not only brought
down into the bosom of the earth, but the earth is
mainly looked at and thought of through the medium
of a purer element. All speaks of tranquillity ; not a
breath of air, no restlessness of insects, and not a moving object perceptible except the clouds gliding in the
depths of the lake." Wordsworth, keenly observant

of all the phases of lake scenery, noticed this effect
as seen upon St. Mary's Loch in the Lowlands of
Scotland.   This sheet of water was frequented in
winter by flights of wild swans which

> " On still St. Mary's Lake
> Float double—swan and shadow."

VII

WELLS AND SPRINGS

" He sendeth the springs into the valleys, which run among the
hills. They give drink to every beast of the field : the wild asses
quench their thirst. By them shall the fowls of the heaven have
their habitation, which sing among the branches. He watereth
the hills from his chambers : the earth is satisfied with the fruit
of thy works."—Psalm civ. 10-13.

THE fluid mineral we call water is all-abounding in
nature. Upon the earth's surface spread forth in wide
expanses, or as rivers, lakes, and bays; diffused through
the air invisibly in its gaseous form, or condensed into
dew, rain, or snow; appearing in the crystal state of ice ;
or, underground, circulating darkly through the earth's
own water-pipes, it is transformed by contact with
the rocks, and, prepared for man's refreshment, issues
into the light of day as fountains and springs.

These changing phases, and this incessant movement,
early attracted attention. In the arid Eastern lands,
which were the cradle of civilisation, wells and springs
acquired an importance unknown to the dwellers in
Western Europe, where the rainfall is abundant and
evenly distributed through the year, and the torture of
thirst all but unknown.

In the *Odyssey* the fountains of Arethusa, where
"sable water glides," are praised ; and the Castalian
springs on Mount Parnassus, sacred to Apollo and the
Muses, and from which the Delphic Pythoness drew

inspiration, are familiar in classical story. The Bible has numerous references to wells and fountains which, whilst supplying man's physical necessities, were regarded as emblematic of the higher spiritual influences essential for his full development. The Israelites in their desert journey, gladdened by the presence of water, sang a song in its praise; and the influence of a righteous man upon his fellows is compared to the vivifying effects of a stream in the landscape: "Who passing through the valley of Baca maketh it a well: the rain also filleth the pools." In the Canticles the graces of the Church are set forth under the similitudes of a "spring shut up, a fountain sealed: Thy plants are an orchard of pomegranates, with pleasant fruits; samphire and spikenard, spikenard and saffron; calamus and cinnamon, with all trees of frankincense, myrrh, and aloes, with all the chief spices; a fountain of gardens, a well of living waters, and streams from Lebanon." And in the well-known conversation between the Master and the woman of Samaria at Jacob's Well, the unsatisfying nature of the material water of which the patriarch drank, "and his children and his cattle," is contrasted with the "well of water springing up into everlasting life." In Bible lands drinkable water was obtained at considerable trouble;[1] even a cup of water had a distinct value—"Whosoever shall give to drink unto one of these little ones a cup of cold water only in the name of a disciple, verily I say unto you he shall in no wise lose his reward." *Early notions.*

Jacob's Well.

Great value of water in the East.

The origin of ordinary springs is simple enough. The sun, which is the great worker of nature, is primarily their cause. The solar heat is constantly raising up water from the land, and from the sea, quietly, but on a large scale, into the vaporous or ethereal form. As the process is silent, and the result *The origin of springs.*

---

[1] Not a valueless gift as it would be in England, but a distinct if small service. Such allusions as these largely lose their meaning as read in this country.

at first not apparent, this action is apt to be under-estimated. An ordinary-sized room, say 15 feet by 15 feet by 8 feet, at an average temperature, may con-tain a pint of water; and, over an acre of surface, a quarter of a million gallons of water may be suspended. This enormous quantity of water vapour rising into the higher regions of the air becomes condensed into rain by the cold currents circulating there, and falls to the earth. The rainfall is distributed in three ways : some of it goes back by evaporation ; some of it drains away into the rivers ; and the residue penetrates the soil, descending until it meets with an impervious stratum of clay or slate, and, finding a vent, appears as a spring. The infiltration of the ocean waters may also cause springs ; and, in plains, water will rise, to some extent, through porous strata, such as gravel and sand, by capillary attraction.[1]

The special character of the spring or fountain will depend upon varying physical conditions, so that the variety produced is considerable.

Perennial springs.

One of the principal types is the perennial spring. Such a spring is fed from a very large body of sub-terranean water, and during long-continued drought does not fail. In the south of France are fountains of remarkable volume ; and in England, where the rain is constant (there being in this country nearly as many wet as fine days), there are some fine springs of this kind. The best-known perennial spring is St. Winifred's Well at Holywell, in Flintshire. The waters gush forth

St. Wini-fred's Well.

where, according to the tradition, Saint Winifred was beheaded by Prince Caradoc. The spring is at the junction of the mountain limestone with the overlying coal measures. This spring never freezes, and has not been known to fail, although it suffers reduction in a very dry season. Over twenty tons of water are

---

[1] A piece of table salt cut from a block and fashioned into the form of a cone may be placed with the base of the cone in a plate of ink to illustrate capillary attraction.

thrown up per minute, forming a stream which actuates a number of mills. The waters proceed from a rock into an artificial basin, over which a building has been erected (the windows of which are painted with illustrative scenes from the life of the virgin saint), and daily devotional services are now held in it.

The volume of water here supplied is surpassed by the fountains of Petrarch at Vaucluse, which, issuing from a cavern at the base of a range of limestone hills, discharge, when least active, 5000 cubic feet or over a hundred tons per minute, forming at once a stream of importance. Upon the melting of the snows the yield is increased, and the annual output is reckoned at 530 million cubic feet. Such a supply involves the presence of a subterranean river, or even the junction of several rivers ; and it is ascertained that certain streams—as the Mole in the Thames system, and the Hamps and the Manifold in the basin of the Trent— disappear into the earth, and reappear as springs. This explains why, occasionally, fish and shells and fragments of plants are thrown out by springs. The sandstone tracts of England are important storehouses of water that might constitute, if necessary, perennial springs of importance. Towns like Manchester, Liverpool, Birmingham, Nottingham, and Derby, standing upon such formations, have an underground source of water in reserve should ordinary supplies fail. Leicester was assisted from this source during the water famine of 1894.

Many springs are irregular in their flow. In some cases this can be explained by the operation of natural forces. Of springs of this kind the younger Pliny described a well near Lake Como which rose and fell thrice daily. The Pool of Siloam (Fig. 70), mentioned in the New Testament, was also another example. It was fed by streams from Mount Zion, and the popular idea amongst the Jews was that the moving of the waters was caused by supernatural

Recipro-
cating
springs.

The Pool
of Siloam.

N

agency—that an angel came down and troubled the

FIG. 70.—Pool of Siloam.

waters, and that it was whilst they were in movement that they acquired their special healing powers.

Modern travellers still report the rising and falling of
the waters.   Other wells of the reciprocating type are
found at Lay Well, near Torbay, and at Giggleswick
in Yorkshire, where the movement is extremely irregular both in time and amount of discharge.   In the
Peak of Derbyshire is the ebbing and flowing well   Ebbing and
between Buxton and Castleton, formerly reputed one   flowing
of the wonders of the district; but it now hardly   springs.
attracts the notice of the tourist ; and at Tideswell, a
few miles distant, there was formerly an important

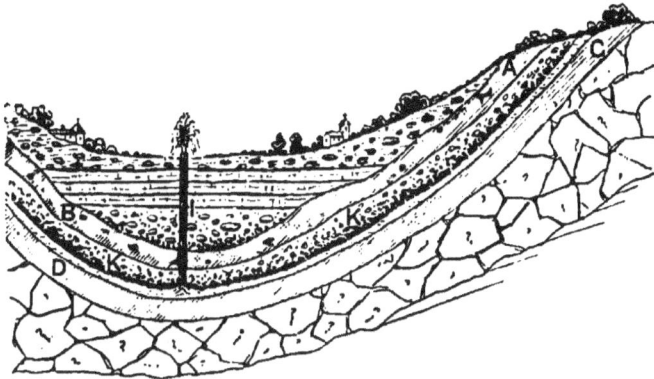

Fig. 71.—Section of Formation for Artesian Well.

well of this kind, of which nothing now remains but
the name.
    Closely connected with the above are intermittent
springs, whose action is explained upon the principle
of the syphon.   There is here a reservoir connected
with the surface below by a channel, which roughly
forms a syphon.   The waters accumulate during the
wet season, but do not begin to flow until the syphon
is filled to its bend.   Once in motion the action will
not cease, however dry the season may be, until the
cavity is completely exhausted, and there may then be
a long period of rest, notwithstanding the prevalence
of rain.

Artesian
wells.

The Artesian spring derives its name from the province of Artois, where it was originally in use. Special
geological conditions must prevail for the sinking of
these *puits Artésiens* to be successful. The most ordinary
formation is when the country is in the shape of a basin
(Fig. 71), and a permeable stratum, such as sand or
gravel, K K, is sandwiched between two impervious
beds of clay, A B and C D. By such an arrangement,
where the agency of man assists the dispositions of
nature, water can often be obtained in large quantity,
and from a great depth, the rise being independent of
the atmospheric pressure which causes the ascent of
water in the common pump, and determined only by
the property of water of rising to its own level. The
temperature of Artesian well water is frequently
higher[1] than that of the atmosphere into which it
rises, through its contact with deeply-seated and heated
rocks, the temperature rising with depth at a definite
rate. In the London basin are many such springs,
one of which supplies a large brewery, and another
the fountains in Trafalgar Square. If necessary, these
wells could be utilised in many parts of England, and
water obtained at a depth, in many cases, of 400 or
500 feet below the surface.

In thermal springs the waters issue at a temperature above that of the isotherm of the place. In
England the most notable thermal springs are at
Bath, which city has, from the time of the Romans,
been noted as a health resort. The baths are fed
from three springs. The King's Bath, which is supplied with 300 gallons per minute, has a temperature
of 120° Fahr., which is about 70° above the normal.
The waters are clear, nearly colourless, inodorous,
slightly chalybeate, or flavoured with iron, and faintly
saline. These ingredients and the gases with which
the waters are charged impart distinct medicinal proper-

Hot
springs.

[1] The earth's crust becoming hotter at the rate of about 1° Fahr. for
each 60 feet of descent on an average.

ties. They have been described as the "medicated waters of the workmanship of nature." Thermal springs are also found at Buxton, which have a temperature of 82° Fahr., not varying either with the hour of the day

FIG. 72.—St. Anne's Well, Buxton.

or the season of the year (Fig. 72). These waters have also medical uses, and with the thin pure air of the town, the highest in the United Kingdom—some 1000 feet above the sea—combine to work cures in rheumatic affections for which Buxton has long been famed. Mary Queen of Scots was a patient here, and the lines which the unfortunate queen inscribed with a diamond

Springs at Buxton.

upon a window pane of her apartment are made prominent to visitors to this health resort :—

"Buxtona, quae calidae celebraris nomine lymphae,
Forte mihi posthac non adeunda, vale !" [1]

The principal well is named after the patron saint of
the town, St. Anno; and in the ancient church crutches
of people who had been cured by the waters were formerly hung upon the walls as votive offerings. In
Derby-shire springs. other parts of Derbyshire there are also warm springs, as
at Bakewell (temperature 62° Fahr.), Stoney Middleton,
and Matlock. At the last-named town, which is finely
situated in the midst of the Anglo-Saxon Switzerland,
hot and cold springs, many of them so charged with
lime as to form petrifying waters, occur side by side in
a remarkable manner. Hot springs of an important
kind are also found in Japan, in the United States, and
elsewhere.

Geysers   In some parts of the world ebullient or boiling
springs occur. Every one has heard of the geysers, or
gushers, of Iceland. This island is highly volcanic, and
the boiling waters are intimately connected with its
igneous phenomena. The temperature of the Great
Geyser, deep down in the shaft, is far above the ordinary
boiling point of water, viz. 257° Fahr. Another boiling
spring of the district gave 232° Fahr., and another 214°
Fahr. The waters of the geysers are highly charged with
silica, and, passing over small objects, gradually encrust
them with a coating of flint. The inhabitants of the
Boiling springs. island, who are short of fuel, utilise these boiling
springs for cooking. Over an area of about half a mile
square, near the White River, about a score of boiling
springs are found, measuring from a few inches to
several feet in diameter. The two most important are
the Great Geyser and the Stroker, or churn. The

[1] "Buxton, farewell ! no more perhaps my feet
Thy famous tepid streams shall ever greet."

Fig. 73.—Beehive Geyser, Yellowstone Park, Colorado.

pool of the Great Geyser is circular, and is 24 feet
across. Ordinarily the flow is small, the waters forming
a rivulet which flows into the lower ground through a
channel it has formed. Every five or six hours they
boil and gush forth with violence, sending up jets
many feet high; and once in the twenty-four hours
there is a great eruption, during which volumes of
water are projected into the air to the height of 70
or 80 feet, with quantities of steam, which obstructs
the view. The Stroker is smaller in its proportions,
the pool from which it rises being 6 or 8 feet across.
When not ordinarily in action this spring can be forced
to discharge by choking up the channel by which the
waters communicate with the depths below with grass
sods and stones. There is then heard a rumbling noise,
and presently a stream of water is projected into the
air to the height of 60 feet, surrounded by clouds of
steam. "Teasing" the Stroker is resorted to should
the visitor arrive upon the scene when the Great
Geyser is not in action. The discharge is due to the
accumulation of steam. Water, whose level is higher
than that of the discharge pipe, occupies a cavern
connected with the spring; and steam produced in
quantity in the space above the water drives it out,
causing an outburst. The geysers and the volcano,
Mount Heckla, which is in the same region, are heated
by the same source—the subterranean fires of the earth.
Hot or boiling springs are also found amongst the
Alps and the Pyrenees, in South Africa and in North
America (Fig. 73). Greenhouses and manufactories
have also been heated by hot springs; and, where the
winters are extreme, ponds and rivers have been preserved from freezing by their means.

Oil
springs.
    Other springs yield mineral oil, naphtha, petroleum,
and other kinds of rock oil. Certain shales in England
and in America (Fig. 74) yield enormous amounts of
inflammable oil, which furnishes a very cheap illuminant
to the humbler classes, and supplies a material of much

importance in the arts and trades. The storage and
conveyance of rock oil in quantity have given rise to
devastating fires upon the American continent.

All spring water contains mineral matter in solution,
often without at all interfering with its transparency.
In some waters the quantity imparts a decided mineral   Mineral
flavour. The mineral compounds are very various, as   springs.

FIG. 74.—Oil Wells in Pennsylvania. The structure on the left is a derrick
for working the boring tool in making a new well. On the right hand
waggons are being filled from the storage tanks.

carbonate of lime, carbonic and sulphuric acid, iron,
silica, magnesia, alumina, and chloride of sodium
(common salt). A division has been made into : (1)
acidulous, (2) chalybeate, (3) sulphurous, and (4) saline.

Waters containing carbonic acid gas sparkle like
champagne, and have considerable chemical action upon
the rocks through which they circulate. Springs of this
class may be seen at Tunbridge Wells, Pyrmont, and
Seltzer Wells, and springs of this kind may evolve

carbonic acid in dangerous quantities, death occurring should this gas be breathed unadulterated. Although the gas itself is poisonous in the extreme when breathed pure, it, strange to say, imparts a distinctly agreeable and wholesome flavour to the beverages in which it is contained. Chalybeate waters contain oxide of iron in excess, not free, but combined with several salts. Such springs form a red deposit, by which they are at once identified. Iron in the blood is necessary to the preservation of health, and chalybeate water is consequently in much esteem for certain complaints.

Sulphur springs. Sulphur springs contain sulphuretted hydrogen, or sometimes sulphate of lime, the sulphur springs of Harrogate containing the former compound. At Baden, near Vienna, the waters contain sulphate of lime or gypsum, one of them supplying over 40,000 cubic feet of water daily. Sulphuric acid is often found in the spring water of volcanic districts, especially in Central America, one of them, described by Humboldt, being computed to yield the enormous quantity of 40 tons of sulphuric acid and 30 tons of hydrochloric acid daily.

Salt springs. Saline springs are impregnated with brine or with medicinal salts. Brine springs furnish the common salt for manufactures and household uses. Some of these waters contain 25 per cent of their weight of salt, and the supply from many springs is enormous. Droitwich, in Worcester, has for two thousand years enjoyed a reputation for its salt springs. The most valuable spring was, however, discovered only a century and a half ago, when a subterranean river of brine was tapped.

English salt-yielding districts. The waters, which are pumped into boilers to be evaporated, are computed to yield three-quarters of a million bushels of salt annually. But the greatest salt-yielding region is in that part of Cheshire forming the basin of the river Weaver, the towns (sometimes known as the wiches) Northwich, Middlewich, and Nantwich,

Fig. 75. The Dropping Well, Knaresborough.

having the salt industry as their staple trade. One of the beds of salt is 75 feet thick, and another over 100 feet. The brine issuing from these beds contains from 3½ to 6½ per cent of salt. There is a remarkable brine spring in Bavaria, near the town of Kissingen, which, besides containing salt, is charged with carbonic acid gas, which at intervals urges the waters into apparent ebullition, and presents phenomena akin to the geysers of Iceland.

Medicinal springs.

Medicinal springs generally contain sulphate and carbonate of soda, as in the Cheltenham waters, which also yield magnesia, the formation being lias with red marl beneath. A spring near Ashby-de-la-Zouch in Leicestershire contains salts of bromine, which give to it a high medical value. In the valleys of Piedmont some of the springs are rich in iodine. The Carlsbad waters hold both the carbonate and the sulphate of soda. The waters flow abundantly, and the yield of these salts is considerable.

Calcareous springs.

Calcareous springs are saturated with lime in solution, which is deposited upon everything the waters encounter in their flow, forming petrifactions often very grotesque and sometimes very beautiful, as, for example, at Matlock and at the Dropping Well at Knaresborough (Fig. 75). Waters carrying calcareous matter also form the stalactites and stalagmites of limestone caverns, as well as the stalagmitic floors of caves. These have enclosed and hermetically sealed up, as it were, in a casing of marble, objects relating to the life of former ages, and which have furnished evidence of the amplest kind as to the ancient history of the earth. Amongst the Himalayas are caverns of this description; nearer home, in Yorkshire and Devonshire. At Matlock, Castleton, and Buxton are caverns where stalactites and stalagmites may well be studied. Water carrying limestone also forms the material known as travertine in South-Western Italy, and tufa or calc-tufa in Devonshire. The latter, formed by the petrifying

FIG. 76. — Robin Hood's Well.

water flowing over the soil containing sticks and
leaves, etc., often exhibits fantastic forms, with a
pleasing play of light and shade upon the surface, and
is abundantly employed in the district in the ornamentation of gardens.

Siliceous
springs.

Siliceous springs are less common : the waters must
be hot to hold much silica in suspension. When they
cool their silica is deposited, forming petrifactions, as in
the case of calcareous waters. In the Azores are such
springs, and in Iceland, as has been named. In the
last - named country the leaves of trees and grasses,
rushes and peat, are covered in the most delicate and
beautiful fashion with a coating of flint of white colour,
through which the minute structure is distinctly
revealed.

Historic
and holy
wells.

Certain wells are reputed holy. In the island of
St. Kilda are three sacred wells, which were visited by
the peasantry in order to make votive offerings. Near
Inverness is St. Mary's Well, which formerly possessed
a similar reputation. St. Cecilia's Well in Netherdale,
Aberdeenshire, was visited upon May morning, the
people walking round it three times leaving gifts.
Robin Hood has given his name to a picturesque
well at Fountains Abbey (Fig. 76). The waters of St.
Mungo's Well were thought to possess a charm against
fairies when belief in "the little folk" was prevalent
in Scotland. The waters of other wells, as at Cambus-
lang, near Glasgow, were thought to incite to madness.
In Wales, and in Ireland, superstitions gathered around
certain wells and springs ; but owing to the spread of
scientific knowledge they are rapidly dying away, with
the customs they induced, and wells and springs are
now regarded as part of the useful work of nature, and
their good services to mankind are generally appre-

The
custom
of well-
flowering.

ciated in an undemonstrative manner. In the Peak of
Derbyshire, however, the elegant custom of annually
decorating the wells still lingers. At the village of
Tissington, near Dovedale, on Holy Thursday, the

villagers, after attending service at the parish church, form a procession to the principal springs, which have been previously decorated with great taste, singing suitable thanksgiving hymns. In these mountain villages the springs are not unfrequently the sole dependence of their inhabitants for their drinkable water, and should these shrink or fail very great inconvenience is experienced.

The uses of springs have been to some extent already indicated; but there are others of importance. They assist the drainage of a country, helping to form streams of great constancy; and, as drainage becomes more important with increase of population, this service will be increasingly appreciated; and, since they both form and feed rivers, the internal navigation of a country is also promoted by their instrumentality. *Uses of springs.*

The greatest service of springs is, however, in furnishing towns with their water supply. The springs amongst the mountains of North Derbyshire are utilised, along with the rainfall, in the extensive waterworks at Woodhead, reaching for miles through Longdendale, for the supply of the city of Manchester and the surrounding towns, the principal reservoirs being 16 miles distant. And Lake Thirlmere, 100 miles from the city, is now also available for the supply of this populous centre.

Newcastle draws its water supply from a distance of 12 miles, and Liverpool nearly 30. The increasing longevity of the people is largely due to the greater attention now paid to sanitation and domestic and personal cleanliness; and in this improvement the efficient water supplies of our great towns bear an important part. *Water supply.*

"Water thus circulating through the earth," says Professor Ansted, "making its way through strata, running down into its crevices and fissures in one place, and rising through open channels to the surface in another—varying in temperature and in mineral contents—cannot but produce great influence on the rock it traverses. That it is chiefly by such means that metals and minerals of almost all kinds, and in all conditions, have been deposited in mineral veins, there can hardly be a doubt, and it is equally certain that, by such means, deposits once made have often been greatly modified, or alto-gether removed or replaced, atom by atom, by other minerals crystallising naturally in different shapes.

Springs agents of change. Thus it is that strata have become metamorphosed and organic bodies preserved permanently by being sealed up and converted into durable stone. It is by the agency of water in its course through the earth that the work of change on accumulations of every kind is rendered easy and effectual. Water is, indeed, the recognised means by which the various physical forces, or rather the different forms of physical force, can most readily act. It is always present—it makes its way through every rock—it forms part of every rock—it conveys fresh material to, and removes part of its substance from every rock that it traverses. Water performs the same part in the economy of the world that blood does in the animal frame. Its use, as seen on the surface, is as nothing compared with its value in circulating through air and earth altogether out of sight."

VIII

THE AIR; MAN'S VITAL ELEMENT

"The atmosphere gives life to earth, ocean, lakes, rivers, streams, forests, plants, animals, and men ; in and by the atmosphere everything has its being. It is an ethereal sea reaching over the whole world ; its waves wash the mountains and the valleys ; and we live beneath it, and are penetrated by it. It is the atmosphere which makes its way as a life fluid into our lungs ; which gives impulsion to the frail existence of the new-born babe, and receives the last gasp of the dying man upon the bed of pain."

Flammarion.

THE true nature of the atmosphere and its precise relationship to man have only within the last century been well understood ; and, even in these progressive times of scientific investigation, there are questions as to its functions which are not yet clearly demonstrated.

The ancient teachers were inclined to trust to appearances in nature. The world, as has been previously stated, they thought was the extensive plain it seemed to be ; the sun was regarded as rising and setting as it appears to do.[1]

Man's native element, "the viewless air," which surrounds the earth like a garment (Fig. 77), by its absence of bright coloration, and by its mobility, long prevented the realisation of its material nature, and it is

[1] The scientific student has to adopt the language which is popular —to speak of the rising and setting of the sun and of the corners of the earth.

O

only by observation and experiment that its true physical properties are revealed. In the olden times it was not only classed as a gas, but was associated actually, as well as by the etymology of the word, with ghosts and the world of spirits. Four elements—earth,

Popular ideas as to the atmosphere. water, fire, and air—were spoken of; but modern science shows that the crust of the earth is highly complex; that water is a binary compound; that fire is not material, but only a form or phase of matter, whilst the air we breathe, and which ·is our first form of food, and the vehicle of life, of movement, and of enjoyment, is

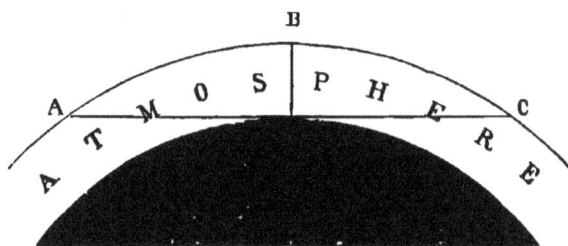

Fig. 77.—The Earth's garment of air.

Diagram showing the influence of the varying thickness of the atmosphere in retarding the sun's heat. A. Line of sun's rays in the morning. B. Line of the rays at noon in the tropics. C. Line of the rays at sunset.

made up of several gases somewhat loosely commingled, and forming what is called a mechanical mixture.[1]

It is difficult to understand how the cultivated Greeks could continually see the effects of air in motion without perceiving its materiality; but as they could not easily grasp the notion of "the impalpable air," or feel its pressure, they included it with the imponderables, and had a very faint idea of its extent and effects.

Pascal's experiment. One of the earliest investigators to dispel the gross ignorance concerning the atmosphere, was the

---

[1] A mechanical mixture, like a cup of tea, contains the sum of the qualities of all its components ; a chemical compound, as water, has qualities unlike those of the elements composing it.

eminent Frenchman Pascal, who has been alluded
to in a previous chapter, and who made his first
experiment upon the subject in the year 1647.
About this time another eminent man of science—

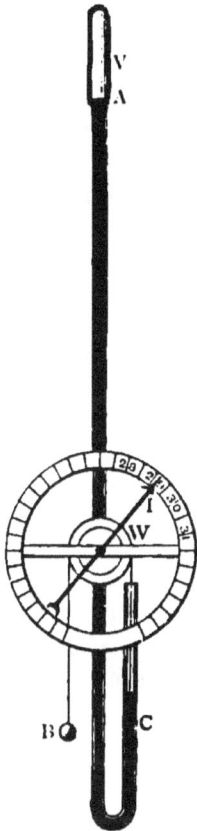

Galileo, a nobleman of Florence—
had been exercised with the fact reported to him that the engineer of
the grand-duke of Florence had found
himself unable to raise water by
means of a common pump to the
height of 50 feet, as instructed to
do; that it was, in fact, impossible
to pump water to a greater height
than 32 feet. The explanation why Galileo's
water rose in the common pump at puzzle.
all was that "nature abhorred a
vacuum," to which Galileo remarked,
half in jest, that this abhorrence
had a definite limit. Torricelli, a
pupil of Galileo, meditating on these
things, showed that mercury could
not be pumped to a greater height
than 29 inches, or thereabouts, and
saw that the explanation given was
worthless. Said he: "Very likely
this horror at vacuity is an idle fancy,
a mere philosophical cant which we
take for good coin without understanding it." The experiment was
modified by pumping wine; and at
length it dawned upon the mind of

Fig. 78.

Torricelli that "the diversity of the Torricelli's
elevation of the two different fluids proceeded from generalisa-
the diversity of their weight, and that they were tion.
supported and counterbalanced by a column of air
of the same diameter, reaching to the top of the First ex-
atmosphere." Pascal carried this reasoning farther, periments
and by experiments of the most decisive kind proved upon air.

the truth of his conclusions, which all subsequent research has corroborated.   Two tubes of glass were
employed, 40 feet long.   One was filled with water and
the other with wine, when it was found that the water
was supported at the height of 32 feet, and the wine
some 7 inches higher, on account of its lesser specific
gravity.   Pascal now concluded that if taken to a
higher elevation the pressure upon the liquids in the
tubes would be less in proportion to the elevation, and

Pascal's ex-
periments.
when the experiment was tried on the top of a lofty
church in Rouen, and, ultimately, upon the summit of
the Pûy de Dome, one of the Auvergne Mountains, the
results were found to be in strict accordance with
anticipation.   The materiality, pressure, and weight of
air were thus proven, and these properties are continually exemplified in the action of the barometer
(Fig. 78).[1]   These early experiments were soon followed
by other researches into the physical and chemical
nature of the atmosphere.

The materiality of the air can be easily proved
without recourse to the air-pump or any elaborate
apparatus.   A miniature diving-bell may be made by
pressing an inverted tumbler into a bowl of water.
Something prevents the entrance of the water, except
to a very slight extent; and that something is the air,
which cannot escape, and, occupying space, prevents another material being in the same place at the same time.

Simple ex-
periments
proving
materi-
ality and
pressure of
the air.
Again, let a wine-glass be filled with water, and
carefully covered with a circular piece of paper rather
larger than the top of the glass.   After allowing the
paper to soak for a minute or two the wine-glass may
be inverted by placing the hand upon it, and it will be
seen that the air will support the water in the glass,
while the paper is bent upwards by the air, which, with
powerful but invisible touch, exerts an upward force

[1] B and C are balancing weights ; W the axis of the wheel over
which the cord moves ; the figures 28-31 in the circumference of the
dial give the range of pressure, the index marking the average for
England ; and V A indicates a Torricellian vacuum.

greater than the weight of water in the wine-glass; the wine-glass might, in fact, be ten yards high without the water escaping. Another easily-performed experiment illustrating the pressure of the atmosphere is to ignite a little dry paper in a wine-glass, placing the hand upon the top when the paper is nearly burnt through, when it will continue to burn for a short time, using up the air, whilst the hand prevents its entering. There will thus be a disturbance of the equilibrium, and the glass will be pressed to the hand, after the air has cooled, with considerable force if the experiment has been properly performed. The writer has frequently tried this experiment without scorching the hand. A very soft hand would show the impression of the rim of the glass, whilst a hard hand might fail, by not thoroughly excluding the outer air. The slightest interference with the even pressure of the air thus causes marked effects.

The actual weight of the air, apart from its pressure, is greater than is generally supposed. A cubic foot, say the quantity that might be contained in a hat-box, weighs an ounce and one-fifth. The air in an ordinary domestic apartment[1] would be computed by the hundred-weight, and in a large hall of a public building by the ton. The whole atmosphere overhead would, in its ordinary state, counterpoise a quantity of the heavy fluid metal mercury to the height of 29 inches over the same area. This is at the rate of 20,000 lbs., or nearly ten tons, upon every square yard of surface of the earth, which will give an idea of the enormous weight which this very light form of matter possesses when aggregated, and may assist the conception that a slight change of atmospheric pressure will produce important changes upon the life and health of man and upon the industries in which he is engaged. The navigator, the miner, the farmer are closely concerned with the alteration in atmospheric pressure; and, at

Weighing the air.

[1] In an apartment 10 feet high, 15 long, and 15 broad, the contained air would weigh a hundredweight and a half.

some seasons, the state of the barometer is as important as the hour of the day.[1]

But the density or weight of the air, as well as its pressure, varies with elevation. The calculations just made are for the sea-level, and might be taken as pretty accurate for most of the English plains. A more permanent impression of the weight of air would be produced by extending such calculations—by employing

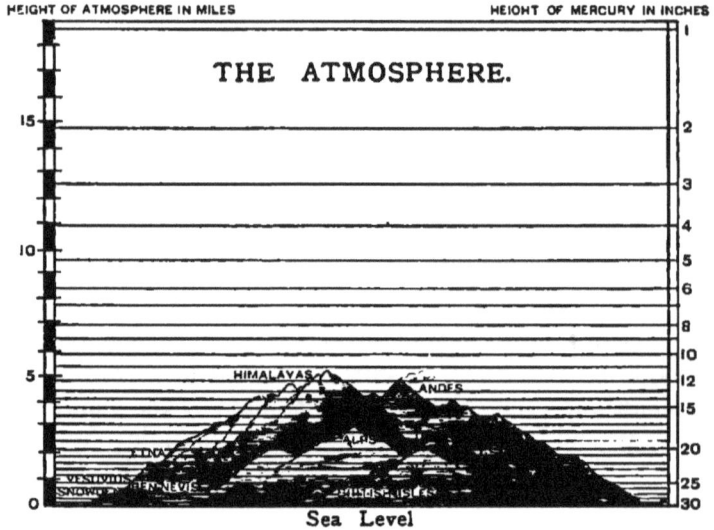

HEIGHT OF ATMOSPHERE IN MILES   HEIGHT OF MERCURY IN INCHES

THE ATMOSPHERE.

Sea Level

FIG. 79.

the dimensions of apartments, for example, with which
Stratifica- the reader is familiar. The air is really stratified, each
tion of the layer from the surface upwards having a different
air. density, and varying in purity, in the amount of water vapour carried in a given unit of capacity, and in other ways. Ascending to higher ground the air (which is closely pressed at the sea-level by the whole mass above

[1] An alteration of barometric pressure of an inch is equivalent to the addition or reduction, as the case may be, of 70 lbs. upon every square foot. It is not uncommon for the press to warn colliers of alterations of 50 lbs. pressure per square foot occurring in 24 hours.

it, just as in a pile of books the lowermost one would
bear the greatest pressure) speedily becomes thinner
and lighter. At the height of 7 miles, a point once   Rare air
attained by Messrs. Coxwell and Glaisher in one of   of upper
their scientific balloon ascents, the barometer registered   regions.
only 7 inches, against 29·92 inches at the surface,
or about one-fourth of the total amount, three-fourths
of the whole mass of the atmospheric envelope being
left below the balloon. At smaller elevations the
pressure is proportionately less (Fig. 79), and so regular
is the decrease of atmospheric pressure with increase
of height that the indications of the barometer,
as in ascending a mountain—for which purpose the
portable aneroid barometer is convenient — may be
taken (with a correction for the reading of the standard
barometer at the sea-line) to mark the elevation.
Thus with the ocean-level reading of 29·92 inches a   Relation of
barometer in the monastery of St. Bernard, at the   pressure to
height of 8130 feet, would read 22·17 ; upon the   elevation.
summit of Etna, 10,893 feet, 20·08 inches ; whilst on
the summit of Mont Blanc, 15,748 feet, the reading
would be reduced to 16·69 inches. Roundly it may
be put that, at the elevation of 10,000 feet, or 2
miles, a third of the whole mass of air is below the
observer ; and that at the height of 18,000 feet,
or 3½ miles overhead, no less than half of the whole
mass of air is below the observer's feet. Not only
does the air become rare with elevation, inducing
more frequent breathing ; it also loses at a rapid
rate the water vapour with which it is charged in the
lower strata.

The presence of water vapour in the atmosphere, if   Humidity
not in excess, is conducive to health. The aeronaut, as   of the air.
he reaches great altitudes, finds the air becoming unusually and disagreeably dry, inducing headache.   If at
14° Fahr. a cubic foot of air can carry a single grain of
water vapour, its power to absorb water will be found to
increase with rise of temperature at a definite rate ; at

30° Fahr. two grains of water vapour may be held; at 40° Fahr., three; at 49° Fahr., four; at 56° Fahr., which is on the warm side of the average outdoor temperature of England, five grains may be contained in suspension. This alteration in physical character is extremely important to health; and in choosing a place of residence, or in selecting a holiday resort, too much care can hardly be given to ascertain the character of the air of the neighbourhood in these and other particulars.

Elasticity of the air.   Another characteristic property of air is its wonderful elasticity. It may be compressed into a small space, and when the compression is removed it will expand again. In the air-gun the compression is carried far, and the spring of the air projects the bullet with tremendous velocity. Most elastic materials lose their power to some extent with use. The bow kept bent becomes less effective, and all springs performing work should, as far as possible, be released when not in action. But the air resembles other gases in this respect, as its spring is not diminished with use, and an air-gun, if quite air-tight, may remain charged for years without injury to its efficiency.

Mobility of the air.   The mobility or capability of being easily thrust aside is another property of the atmosphere which is of the utmost importance to the life and convenience of man and animals. It is brushed aside by an insect's wing vibrating many times in a second, and the infant walks unimpeded through the air, his native element. Its particles easily moving over each other are parted by the slightest effort, and progress through the air is not only easy but pleasant.

Medium of light, sound, and perfumes.   As the medium of light, and colour, of perfume, and of sounds, the atmosphere ministers to the needs and to the enjoyment of men. Without the air there would be no twilight, no half shades—all would be glare or gloom. The sun would appear above the eastern horizon unannounced by the dawn, and, his daily course being run, would sink suddenly without warning

in the west. Like water or glass the air bends aside Refractive
the rays of light traversing it, and beautiful and power of
useful effects are thereby produced. This refractive the air.
property of the invisible air may be illustrated by the
action of water in an experiment which all may try. If
in a toilet bowl a dark-looking object, such as a bronze
medal, or failing that a penny, be placed in the centre,
and attached to the bottom with a little wax, on retiring
from the bowl until the coin is just concealed by the
edge of the vessel, water poured into the bowl by
a confederate will bring the coin into view. Although Experi-
neither the coin nor the spectator has moved, the water ment upon
turns aside the rays of light proceeding from the coin refraction.
and shows it in a place where it does not really exist.
The air surrounding the earth deflects the light from
the sun, the moon, and all the heavenly bodies. The sun,
it can be shown, is seen before it rises—that is, before
it is actually or geometrically above the horizon—and,
in the evening, in these latitudes, we often see it for
about an hour after it has really set, owing to this
curious action of the atmosphere.

Perfumes and sounds travel through the air with
great ease; not always with the same readiness, but with
wonderful facility. A grain of musk in an apartment
will give off its fragrance for weeks without apparent
diminution of its size, so subtle is the essence, and so
easily is it diffused through the air. Without the air
there can be no sounds. When it is thin and rare, as
upon a tall mountain, sounds are diminished. A pistol
fired upon the top of Mont Blanc gives but a feeble
report; and when a bell is rung in a vessel from which
the air has been pumped there is no sound whatever,
however violent the action may be.

The chemical properties of the air are not less signifi- Chemical
cant. This part of the subject has been thoroughly properties
worked out during the present century, and the last ten of the air.
years have added further information of much value. It
was with surprise that people, in the year 1774, heard,

through Dr. Priestley, that the light, fine, and invisible air was not simple and uniform but of a composite nature. It was shown that one of the gases composing common air was oxygen, or "dephlogisticated air," as it was at first named. The following year a Swedish chemist eliminated oxygen from air, naming it "empyreal air" from its property of supporting flame. Another experimentalist, an Englishman of rank, engaged also in this

Cavendish's labours.

investigation. This was the Hon. Henry Cavendish, whose experiment of "weighing the world" was described in an early chapter of this series. Cavendish is sometimes styled "the father of pneumatic chemistry," and it is certain that the experiments he performed were conclusive and satisfactory. As the life and labours of this eminent chemist are but little known, except in scientific circles, it may be interesting to say that, endowed with immense wealth, he preferred to live a very simple life, and, neglecting his large social opportunities, to devote his energies and his means to the prosecution of his favourite studies ; and with the greatest success, for no problem seemed too abstruse for Cavendish to attack. So earnestly did he devote himself to this work that the duties of citizenship and friendship were sometimes neglected, and his biographer relates some curious stories

Eccentricities of a scientific man.

of his eccentricities : "His town residence was close to the British Museum. Few visitors were admitted, but some found their way across the threshold and have reported that books and apparatus formed its chief furniture. For the former, however, Cavendish set apart a separate mansion in Dean Street, Soho. Here he had collected a large and carefully-chosen library of works on science, which he threw open to all engaged in research, and to this house he went for his own books as one would go to a circulating library, signing a formal receipt for such of the volumes as he took with him." His favourite residence was a beautiful suburban villa at Clapham, which was occupied by workshops and a laboratory. "It was stuck about with thermometers,

rain-gauges, etc., one of the former being of his own
construction. The upper apartments formed an astronomical observatory ; one room was used as a forge, and
upon the lawn was a wooden stage for mounting a large
tree, the top of which was a favourite station for
astronomical, meteorological, electrical, and other researches." Here Cavendish lived comfortably, but
making no display. His few guests were treated on all
occasions to the same fare, and it was not very sumptuous;
if any one dined with Cavendish, he invariably gave
them a leg of mutton and nothing else. "On one
occasion three or four scientific men were to dine with
him, and when his housekeeper came to ask what was
to be got for dinner, he said : 'A leg of mutton.' 'Sir,
that will not be enough for five.' 'Well then, get two,'
was the reply." But this arose simply from the wish to
save himself trouble ; for upon occasion he could be
liberal enough, as indicated by the following anecdote :
" His library being out of order, a gentleman, who was
not very well off, was called upon to classify and arrange
it. The work occupied a considerable time, and ultimately the gentleman left. Mr. Cavendish, being at
dinner one day at the Royal Society, some person
present mentioned this gentleman's name, upon which
Cavendish said : 'Ah! poor fellow, how does he do?
how does he get on ?' 'I fear very indifferently,'
said this person. 'I am sorry for it,' said Cavendish.
'We had hopes you would have done something for
him, sir.' 'Me, me, me, what could I do ?' 'A little
annuity for his life, he is not in the best of health.'
'Well, well, a cheque for ten thousand pounds, would
that do ?' 'Oh, sir, more than sufficient, more than
sufficient !'" Other acts of liberality are recorded of
this extraordinary man, to whom no problem in physics
or chemistry came amiss, and who is justly ranked as
one of England's very greatest natural philosophers, but
who, like Sir Isaac Newton, John Dalton, and others
that might be named, concentrated their intellectual

energies upon their scientific work to the danger of neglecting the little things of life. As a scientific worker his methods were orderly in the extreme,[1] and his ideal was lofty, and was pursued in the most complete manner. "The pursuit of truth was with him a necessity and a passion. In all his researches he displayed the greatest caution, not from hesitation or timidity, but from a recognition of the difficulties which attend the investigation of nature, and from his delight in reducing everything to numerical rule, and his hatred of error as a transgression of law. *Cavendo tutus,* the motto of his family, seems to have been constantly before him."

Composition of the air.
The researches of these investigators show that common air is made up of oxygen and nitrogen as the chief components, with carbonic acid, water vapour, and other gases in very small proportions. According to Lavoisier, the composition, as published in 1777, was—

	Weight.	Volume
Oxygen (O) . . .	23	20·8
Nitrogen (N) . . .	77	79·2
	100	100·0

Subsequent researches show that the air by volume contains—

Oxygen	20·61
Nitrogen	77·95
Carbon dioxide . . .	·04
Water vapour . . .	1·40 (varying)
Nitric acid ⎫	
Ammonia ⎬ traces	
Carburetted hydrogen ⎭	
In towns—	
Sulphuretted hydrogen ⎫ traces	
Sulphuric acid ⎭	
	100·00

---

[1] Sometimes carried to excess, as the visits of his tailor were marked upon the almanac and not deviated from, and "his walking-stick was always placed in one of his boots, and always in the same one."

Of over 500 analyses of town and country air in winter and summer, in wet and in clear weather, the result invariably came out that, of 100 volumes, oxygen was 20·833, and nitrogen 79·167. This analysis takes no note of carbonic acid gas, which is in variable quantity, or of water vapour, also a variable quantity, and other gases, which, although in small proportion, are nevertheless of importance. The atmosphere is a mechanical mixture—that is, the principal components are blended without losing their special properties in consequence. Other proofs that it is not a stable chemical compound like water, or salt for example, are derived from the chemistry of the subject.

Of the existence of water vapour in the air there is indubitable proof,[1] although it cannot be seen ; it is even misleading, as when in large quantity it may cause the air to be unusually transparent. The amount it does or can contain depends upon its temperature, upon the nature of the winds blowing at the time, and upon the state in which the water exists in the air. The quantity of aqueous vapour in the air is, in this country, considerable ; but in some parts of the interior of Asia, Russia, and Africa the air is sometimes remarkably dry. A piece of sea-weed roughly indicates the humidity of the air, and rock salt also becomes dry or damp according to the amount of water vapour present. The quantity of carbonic acid, or carbon dioxide, in the air, is small and not quite constant. One of the earliest determinations showed that of 10,000 volumes of air this gas formed 4 volumes, and subsequent researches by French chemists, and in England by Angus Smith and Roscoe, confirm this result. **Water vapour in the air.**

The oxygen in the air is the active principle. "Busybody" oxygen, as it has been called, unites with nearly every element, and it supports respiration and combustion. It is invisible, is rather heavier than air, and **Busy oxygen.**

[1] Clothing unused becomes damp, and a cup half full of sulphuric acid will grow in volume by attracting water vapour from the air.

Nitrogen.

is as abundant in nature as it is energetic, something like half the materials around us being composed of this gas. Nitrogen is the diluting principle in the air, of which it forms four-fifths by volume. It puts the curb upon the activity of oxygen, and prevents the processes of life from being too energetic. Its properties are negative ; it will not burn or support combustion ; and, in an atmosphere of nitrogen, a person would be suffocated as by water, not because it is poisonous, but on account of being unable to respire. It is lighter than air in the proportion of ·9748 to 1.

Carbonic dioxide.

The carbon dioxide contained in common air is not quite constant. During the night, and also during dry winds and fogs, the proportion increases, sometimes to double the average percentage. In country districts and also over the sea it is diminished. This heavy and poisonous gas (specific gravity 1·5123, air taken as 1) is necessary to the growth of plants; and with regard to human life, although poisonous in the extreme to breathe, it pleasantly dilutes effervescing beverages, for when taken into the stomach its properties are wholesome rather than injurious. It sometimes accumulates in dangerous quantities in wine cellars, vaults, and deep wells, where it is given out.

Minor components of the atmosphere. Ammonia.

Ammonia is also found in the atmosphere, but in extremely small and variable quantity. It is absorbed by the rain passing through the atmosphere ; and the rain water of London has been found to contain 3·45 parts of ammonia, Liverpool 5·38, Manchester 6·47, and Glasgow 9·10 parts per million. Insignificant as the amount is, the effect upon vegetation of this component is very marked.

There are also various impurities in the air, as saline and organic matter. Sulphates are found in quantity in the air of towns, and there are also micro-organisms existing in enormous numbers, whose effect upon the atmosphere is being further investigated.

The variations in the temperature, the contained Weather.
moisture, the electrical state, and the movements of the
air determine the weather and climate of a place. In no
country is the state of the weather of more importance
than in England, and the daily press, recognising this
fact, gives diagrams showing the leading weather facts
(Figs. 80, 81). In countries where the weather is fixed
and constant some other physical influence may be dominant. In one region a great river is all-important; in
another part of the earth, frost and snow limit and

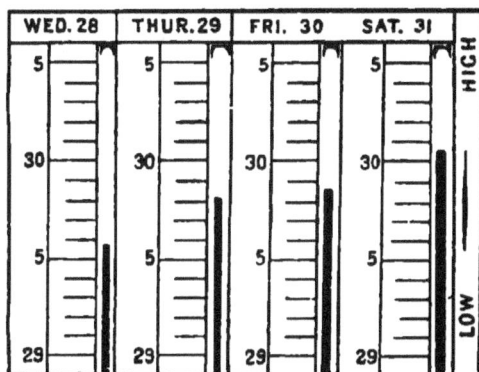

Corrected to sea-level, and reduced to 32°F.

Fio. 80.—Specimen of *Daily News* barometer chart.

bound the operations of man. The weather in England is constantly changing, and so cognisant are
its people of this fact, that they excite the criticism
of foreigners by the persistent manner with which
they remark upon its endless permutations. It has
been urged in defence that the English show their
practical sagacity by recognising the supremacy of the
weather.

The word weather is from the same etymology
as the word wind (A.-S. *weder* = wind), and here
again a correct insight is shown as to the nature
of things, for the wind seems to control the other

weather factors—elevation, temperature, moisture, etc. So much is this the case that it has been remarked the wind *is* the weather. The winds of England are chiefly westerly. The country is not only surrounded by the sea ; it is also governed by the strong winds from the

Direction of the winds.

wide and deep Atlantic. The mean direction is south 16° west. For half the year, upon an average, the winds have an element of west in them, and the westerly type of weather, giving cool summers and mild winters, is the normal English condition ; and as such winds are

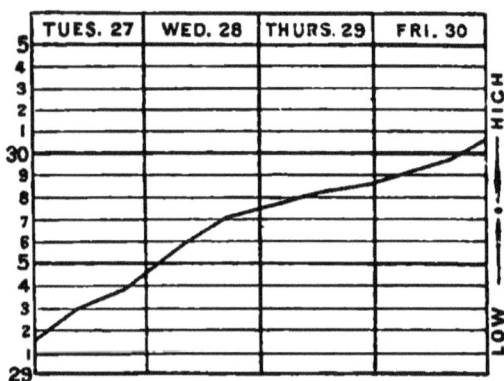

Fig. 81.—Specimen from *Daily Telegraph* weather chart, showing range of barometer.

often stormy, the influences of the west are thus increased.

Velocity of the wind.

The speed as well as direction of the wind is important. In England the air is calm on about thirty days in the year upon an average. The climate is distinctly windy. A candle, for example, can rarely be carried unshaded in the open. A light wind will travel some 4 miles an hour ; a moderate wind 12 ; a breeze 17 ; a gale 45 to 50 ; a violent tempest from 80 to 100 ; and a cyclone will touch 150 miles an hour.

In this country there is no distinctly injurious wind,

the nearest approach being the east winds which are
prevalent during the spring, and which have acquired
an evil reputation ; but we have nothing approaching
in inconvenience the local winds in some countries
abroad, as the " bise " of North and Eastern France, a
north wind, cold, and violent ; or the same wind felt in
North Italy and other parts of South Europe, where it
is known as the " bora." In South and South-Eastern
France the mistral, an extremely cold wind from the Local
Alps and the Pyrenees, suddenly sends down the tem- winds.
perature, and for a time reverses the climate. Another
local wind is the harmattan, which prevails on the
north-western coast of Africa, between Cape Verde and
Cape Lopez, during December, January, and February,
blowing from the desert. It is extremely dry, causing
the lips to crack, but otherwise regarded as healthy,
showing that the rains are over. The blizzard of the The
Northern States of America and of Canada is also a blizzard.
strong driving north wind, before whose influence man
and his works are impotent. The blizzard often rises
very suddenly, and whoever is overtaken by it far from
cover is in the greatest peril. A wind of an opposite
kind is the simoom of the African Desert, or of India,
where it is called "the devil wind." In the Sahara its
power is supreme, hot as from the mouth of an oven,
and charged with fine dust and sand ; but fortunately
of no long duration. Caravans crossing the desert are
often overcome ; in one case 2000 men and 1800 camels
perished in the simoom. There are reports of armies
being lost through this wind, as the one of 50,000
soldiers sent by Cambyses to destroy the temple of
Jupiter Ammon. The terrible destruction wrought by
the simoom is also shown by the bones of men and
camels that lie bleaching upon the caravan routes.

In this way the vital fluid of man is converted into
his destruction. These are, however, exceptional workings. The atmosphere, on the whole, admirably
serves its purpose. Its properties might be modified

P

with good effect in certain cases ; but not without interfering with its general suitability for the business of
life. Closer examination of its physical, chemical, and
electrical state may indicate how to take fuller advantage of its useful properties, and how any dangers
arising from it may be avoided.

IX

THE WINDS OF HEAVEN

" I've swept o'er the mountain, the forest, and fell ;
I've played on the rock where the wild chamois dwell.
I have tracked the desert so dreary and rude,
Through the pathless depths of its solitude ;
Through the ocean caves of the stormy sea,
My spirit has wandered at midnight free.
I have slept in the lily's fragrant bell ;
I have moaned on the ear through the rosy shell.
I have roamed along by the gurgling stream ;
I have danced at eve with the pale moonbeam.
I have kissed the rose in its blushing pride,
Till my breath the dew from its lips has dried.
I have stolen away on my silken wing
The violet's scent in the early spring.
I have hung over groves where the citron grows,
And the clust'ring bloom of the orange blows.
I have sped the dove on its errand home,
O'er mountain and river, and sun-gilt dome.
I have hushed the babe in its cradled rest,
With my song, to sleep on its mother's breast."

As the land has its rivers, and the ocean its currents, so the air has its movements. These aerial
rivers or winds may be broad, long, and permanent,
like the great equatorial flow of the Pacific, or narrow,
deep, and fitful; and they move horizontally, or

Gods of the winds. descend or ascend a gradient. Impalpable to the touch when at rest, and invisible except when seen in mass (the blue sky being simply the atmosphere coloured by water vapour), the air gathers force when in rapid movement, producing storms and phenomena of the most striking kind. Great momentum may result from a body of light weight, like small shot, moving with high velocity, or from a large mass of matter moving slowly, as when an iceberg of 1000 tons strikes a vessel at sea, or collides with a neighbouring berg, when the impact is tremendous. The motion of air is akin to the first-named case. Air is a light material— still it is a material; and when moving its mechanical effect is considerable, its rate of motion being sometimes over 100 miles an hour.

Primitive notions. In early times the winds were thought of as irregular and lawless, or as under the control of special deities. In the mythology of Southern Europe, Boreas, **Wind gods of the classical mythology.** who caused the north wind, was supposed to dwell in a cave in Thrace, and was worshipped by the Athenians ; and Zephyrus, the personification of the west wind, is often mentioned by Homer. When a higher form of religious belief prevailed, the winds were still regarded as erratic and uncontrolled. " The wind bloweth where it listeth, and thou hearest the sound thereof, but canst not tell whence it cometh nor whither it goeth," [1] said the Master in His memorable interview with Nicodemus ; and centuries passed before it began to be understood that the winds were subservient to physical law and order—that their action was not erratic ; that they could be classified and systematised, and that in some parts of the earth their movements were as regular as the rising and setting of the sun.

The wind sprite. In the Scandinavian mythology the same idea of separate deities controlling the various workings of nature prevailed as amongst the people of Southern

[1] John iii. 8.

Europe,[1] and until comparatively recent times lingered the belief in the water sprite and the wind sprite, whose favour had to be propitiated by presents and by ceremonies. The relics of such ideas still remain in our language and leaven our poetry and proverbial speech.[2] The causes of winds could not be fully understood until astronomical, physical, and general science were in a somewhat advanced state. The gases composing air do not stratify according to their relative density, even when the air is at rest; but the thorough mixture of these components is greatly promoted by the motion of the air we call wind.

The primary cause of all winds is the varying heat of the sun. Equatorial regions have vertical sunshine where the average temperature of the air equals 80° to 85° Fahr. Where the incidence of sunlight is oblique, as in the temperate and polar zones, the result is less,[3] the average temperature of London being about 50° Fahr. ; and in higher latitudes the heating is proportionately diminished. Air, like other gases, is very sensitive to the action of heat. It expands quickly and to a large degree when heated, and contracts with the same

Causes of winds.

---

[1] Vide *Feats on the Fiord,* by Miss Martineau.

[2] An interesting illustration is furnished in the "Song to Ægir," set to music by the Emperor of Germany, and translated by Professor Max Müller :—

> " Hail, Ægir, Lord of Billows, whom elf and sprite obey,
> To thee in morn's red dawning the hosts of heroes pray ;
> We sail to dread encounter, lead us o'er surf and strand,
> Through raging storms and breakers, into our foemen's land.
> Should water-demons threaten, or should our bucklers fail
> Before thy lightning glances, make thou our foemen quail ;
> As Frithjof, on Ellida, crossed safely o'er the sea,
> On this our dragon shield us, thy sons who call on thee.
> When hauberk rings on hauberk in battle's furious chase,
> And when the dreaded Valkeries our stricken foes embrace,
> Then may our songs go sounding, like a storm-blast o'er the sea,
> With clash of swords and bucklers, thou mighty lord, to thee."

[3] At the tropics the air is expanded by the solar heat, and is, therefore, lighter than in temperate or polar regions. If taken at 758 millimetres in the tropics, 761 might represent the mean pressure in the temperate zones, a difference of a tenth of an inch—an amount quite sufficient to cause strong winds.

facility when cooled. The influence of the sun may be easily imitated, so as to give a mimic representation of the winds. The domestic fire in an ordinary apartment acts on a small scale the part of the great earth-fire, the sun. In the chimney there is an ascending current of hot air; and the heated air at the equator, which may be thought of as the earth's chimney, constantly rises. If the doors of a room be closed, and a lighted taper be placed at the bottom, the inrush of cold air to supply the vacancy caused by the air leaving by the chimney will be perceived; the flame will be blown inwards and perhaps extinguished. This may be taken to typify the movements of cold air towards the earth's chimney, or the equator, from each side to supply the loss caused by the air there ascending. The rotation of the earth is also a modifying cause of the winds. The varying amount of vapour under different circumstances is also to be taken into account. It is the vapour in the atmosphere that retains the heat of the sun, dry air being transparent for heat; and the atmosphere also acquires its warmth from contact with the warm earth by conduction, so that where the vapour tension is greatest there the greatest warmth will be engendered.

Winds may be classified into constant, periodical, variable, and storm winds, and a map may be constructed of the great aerial movements, just as in the case of ocean currents.

The steady inflow from the regions north and south of the equator of cool air near the surface—and it must be borne in mind that the air current nearest the surface is to be understood unless otherwise specified (every such current having an upper one in the opposite direction)—gives rise to the trade winds. The original derivation of the name of these winds, which, it may be remembered, were discovered by Columbus in his voyage across the Atlantic, was not, as is popularly supposed, from their usefulness to

Simple illustration of wind.

The trade winds.

THE WINDS OF HEAVEN 215

commerce, as the name was originated before it was
ascertained that they would assist navigation. To the
older writers who were seafaring men, the word trade
was synonymous with track or course—that is, the winds
whose course or trend was constant and fixed; but
it will be difficult to dissociate these winds from the
idea of commerce, especially as they have helped for
centuries to convey sailing vessels from the Old World
to the New. The sailors of Columbus were at first
pleased to find a steady breeze carrying them the way
they desired to go; but when it·was found that for
days and weeks the wind was blowing from the east,
it was forced upon them that there would be difficulty
in returning to Europe, and they grew, as has been
related, discontented and mutinous. In the Atlantic
and Pacific the trades prevail from 10° to 28° north
of the equator, and from 5° to 28° south. They are
thus associated generally with the tropics, swinging
a little with the season, the band of the trades
travelling northwards with the sun at the summer
solstice and southwards again with the sun, increasing in southward declination till the winter solstice is
reached. The trade wind of the southern hemisphere
is better defined than that blowing north of the
equinoctial line. The velocity varies from 10 to 20
miles an hour, and the trades may variously be described as a brisk wind increasing to a breeze. Navigation is rendered extremely easy, a ship travelling with
the trades 100 to 150 miles a day, the mariner
scarcely touching the sails. In strong contrast with
the stormy seas around the Spanish peninsula—notably in the Bay of Biscay—is the ocean lying under
the influence of the trade winds, so that the sailors spoke
of this part of the ocean as *el Golpo de las Damas*, "The
Sea of the Ladies." The testimony of travellers is almost
unanimous as to the pleasure experienced in these
latitudes. The atmosphere is transparent, the sky un-
clouded, the winds soft and refreshing without being

Origin of the name.

Limits of the trade winds.

Pleasant sailing.

"The Sea of the Ladies."

cold.   Columbus compared the experience here with
the balmy mornings in April in Andalusia, "wanting
but the song of the nightingale and the sight of the
groves to complete the fancy that he was sailing along
the Guadalquivir." The sunsets are especially fine.
Upon a crimson background light clouds, often in long
streaming lines of purple and gold, are seen, the high
coloration fading gently away into night, when the
stars shine with unusual brilliancy. To some enterprising spirits the constancy becomes tedious. "This
sailing of the trades," says a modern writer, "with one

Monotony of this region. unvarying form of cloud overhead, and that not very
picturesque, and with a constant everlasting blow of
wind, never rising to a storm, never sinking to a breeze,
without lightning and without rain, is somewhat mono-
tonous in itself, and makes an uninteresting time to

The trade winds and naviga- tion. many." The bearing of the trades upon navigation
is shown by the fact that a small fishing smack has
been known to cross the Atlantic assisted by the trade
winds, and the return trades, which constitute the prevailing south-west winds of the English climate, although much less constant, were taken advantage of
for the journey from America to England before steam
power came into general use.

As the winds from the north and the south approach
the equator they gradually become heated and ascend
rather than move along the horizontal, so that near the
line there is a narrow calm belt known as the doldrums.
It is widest in the autumn and narrowest in the winter,
the breadth varying in the Atlantic from 150 to 550
miles.   The sufferings of a ship's crew becalmed for
days, immediately under the equator, have been vividly
described by Coleridge in his story of "The Ancient
Mariner."

If the earth had no rotatory motion the cold winds
from the polar parts and the temperate zones towards
the equator would be due north and south winds; but
they are strongly influenced by this rotation, which

becomes more marked as the equator is approached.
The air being material possesses the general properties
of matter, one of which is that of inertia, or the
tendency to continue in the state of rest or motion in
which it is found, and to resist any change from one
state to another. At the equator, where there is the   Easting of
largest circle of diurnal motion, the eastward velocity   the trades.
is over 1000 miles an hour, and the atmosphere par-
takes of this eastward motion. At the equator the
rotation towards the east is 1042 miles an hour; in the
latitude of 30° north or south it is reduced to 883 miles
an hour. It will be seen upon a moment's reflection
that wind moving from the edge of either tropic towards
the equator would gradually pass over parts of the
earth having a higher rate of easting than itself.
What will be the result? The air current is unable
at once, on account of its inertia, to take up this
increased speed, and lags back in the direction from
which the earth is moving, so that this northerly and
southerly inrush of cold wind is deflected and becomes
a north-easterly wind[1] north of the equator, and a south-
easterly wind south of the equator. A familiar illustration is afforded by the circumstance that a person
stepping sideways on to a moving tram-car is thrown in
the opposite direction to that in which the vehicle is
moving. A succession of such unwise persons, all
possessing this property of inertia, would be similarly
treated, and a stream of air is steadily deflected in the
same way. The difference[2] in the rate of motion
becomes less, say in travelling over ten degrees of
latitude; but, as the inertia is gradually overcome on the

---

[1] A north-east wind is one travelling towards the south-west.
[2] The eastward velocity of the earth, at various latitudes, is
estimated thus :—

At the equator 1042 miles per hour.
10° N. or S.	1026	,,	,,
20°	,,	980	,,	,,
30°	,,	883	,,	,,
40°	,,	800	,,	,,
50°	,,	620	,,	,,

(London, 51½° N.)

journey, the easterly motion becomes more pronounced, until the trades seem to come from almost due east, and in the region of calms they become so heated as to ascend in a column like rising smoke, and their horizontal movement is no longer perceived. Although, as has been shown, the name was given before it could be known that these winds would benefit man in his commercial dealings, it must be acknowledged that it is very appropriate. " It was this constant and gentle wind," says the author of *The Earth and Man*, " that carried the great navigator Magellan, whose ship made the first voyage round the world across this vast ocean, and who gave to it the name of Pacific, which has been preserved to the present day. It is by this line that the Spanish galleons, laden with the gold of the New World, accomplished during more than two centuries their peaceful voyages from Acapulco to Manilla, sheltered at once from tempests and from the attacks of the nations envious of so much wealth." The sky, it may be remarked, is more or less covered with a small detached cumulus cloud of the forms shown in Figs. 110 and 111. " Sometimes," says the Hon. Ralph Abercromby, " a thin, hard, broken strato-cumulus covers the sky with such regularity that, when seen in perspective near the horizon, we look at a series of bars, like the leaves of a Venetian blind ; but if the gradients are at all steep, squalls and showers from cumulus cloud are of frequent occurrence in the wind regions." It has been pointed out that it is not only at sea that these winds are helpful, but that within the tropics inland navigation is also promoted by their agency. Over the lower course of the Amazon there is a prevailing east wind, which enables the voyager to make headway against the stream. Over the basin of the Orinoco east winds are also similarly constant.

Winds constantly moving in a given direction occasion a counter-current of equal importance ; and the return

Marginal notes:

Uses of the trade winds.

Trade winds and inland navigation.

Anti-trades.

trades are amongst the most useful of such winds. These winds, at first an upper current, cool and come down to the surface. The existence of counter-currents is illustrated by observing that in the trade-wind regions clouds at a great elevation may be seen moving towards and not from the east. Humboldt, the eminent Prussian scientific traveller, upon ascending the Peak of Teneriffe (the Canary Islands being at the verge of the trade winds), observed that the easterly wind at the surface of the sea was exchanged at the summit for a strong westerly one. The ashes of volcanoes in eruption in these regions have been projected into the counter-current, and have descended to the east of the point of discharge. The ascending currents of air at the equator, over a band 24,000 miles long and some 47° in breadth (equalling 3000 miles), are cooled as they ascend, and eventually reach their highest point; and, unable longer to move in the vertical, and prevented by the force of the ascensional column at the equator from moving in that direction, find out the path of least resistance, which is northwards and southwards from the equator. These return trades retain the high eastward movement acquired at the equator and steadily descend towards the earth. Striking the surface they overrun it, and an opposite state of things occurs to that described as characterising the trade winds. The earth is moving eastward, but the return trades have a higher eastward motion, outpacing the earth in the direction in which it is travelling, and in effect become westerly winds or winds moving to the east. This action may be illustrated by setting in motion two circles one within the other, revolving in the same direction, but the outer one having a considerably higher speed than the other. If the inner circle be gradually made to approach the outer one until they touch, an object upon the inner circle, which may be taken to represent the earth where the trade winds descend, will be seen to be struck forward in the

Currents and counter-currents.

The return trades.

direction in which it is going, the outer circle represent-
ing the winds.

The winds of England.   It is these return trades or south-west winds which
have such marked effect upon the English climate.   At
London, on an average, the south-west prevails upon
112 days; the north-west, 50; west, 53; or 213 days

SW = 104
NE = 48
N = 41
W = 38
S = 34
NW = 24
E = 22
SE = 20
O = $\frac{34}{365}$

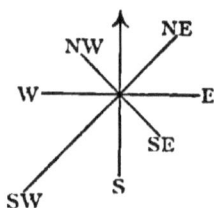

upon which winds of a
more or less westerly type
blow at London.   For the
whole of England the dis-
tribution of the winds is
not very dissimilar (Fig.
82).   The returns also
show that the south-west

FIG. 82.—The Wind Rose for
England.

winds are distributed more
evenly through the months

Westerly winds.   of the year than any other.   The differences owing to
locality are not sufficient to interfere with the general
conclusion that it is the westerly type of weather that
is distinctly English.   In Edinburgh, for example, the
westerly winds blow on 230 days; in Ayrshire, for
three-fourths of the year; and on the east coast of
Scotland, as along the shores of Moray Firth, westerly
winds are reported for two-thirds of the year.

These return trades, possessing a high temperature,
have great carrying power for water vapour, so that
they arrive upon the British coasts laden with moisture,
The warm west winds.   which is abundantly discharged over the land.   These
islands have the reputation of being rainy; the um-
brella is a necessary outdoor accompaniment.   Then as
they travel from the tropics, and chiefly along the
course of the Gulf Stream, a river of sea water broad
and deep, they act the part of a natural warming
English weather.   apparatus for Britain.[1]   The cooling of the Stream,
even on reaching somewhat high latitudes, is very slow,

[1] Warm air passing over parts of the ocean colder than itself loses
its temperature slowly.   An exchange goes on according to a definite
physical law.   If air had the same weight and specific heat as water,

and as far north as the fjords of Norway the sea is kept open and navigable, whilst ports in the same latitude across the Atlantic are sealed up in ice. That the winters of England are mild and damp is due to the return trades. On the other hand, the ordinarily cool English summer is equally a result of the prevailing westerly winds, which have a lower temperature than the land heated by the fervent sun. The winds are the principal weather factor, and our equable climate is due to them—a climate without great extremes, and the finest working climate known, enabling England to become one of the world's great workshops; the climate of which it was remarked by Charles II. that one could be abroad more hours in the day and more days in the year with comfort than in any other known; and this notwithstanding the admitted drawbacks. It is often said that it is to the Gulf Stream and the position of the British Isles upon the western verge of Europe and on the east of the Atlantic that the English climate is indebted; but it must be borne in mind that winds are the primary agent, and that the Gulf Stream and the Atlantic are quite inoperative during a land wind from Europe, and that under the influence of the east or north wind the weather is not of the English type.

The finest working climate.

The characteristic winds of England are also its strong winds. We have on an average only a month in the year of calm weather. The westerly, south-westerly, and north-westerly winds bring storms. The country is well swept, winds being the ventilators of the earth; and the favourable health returns, the

The winds as ventilators.

then 1 cubic mile of sea water, by rising one degree, would cool 1 cubic mile of air passing over it one degree ; but as air is 770 times lighter than water, and has only one-fourth its specific heat, it follows that 770 × 4, or 3080 cubic miles of air, passing over the Atlantic, will lose one degree by 1 cubic mile of water rising one degree. Both the layer of water and that of air must be supposed quite thin, so as to make up in area what they lack in depth. If, on the other hand, the wind travels from a colder quarter than the sea over which it passes, a mile of water falling one degree would raise 3080 miles of air one degree.

physical vigour of the nation, and the high average longevity, are to some extent a consequence of the winds.

The atmospheric pressure over the whole globe is pretty generally thus—(1) Close to the equator a belt of approximately low pressure. In this region are the Atlantic calms or doldrums of the navigator, where the weather is sultry and the air so wanting in horizontal movement "that a ship may lie for weeks on the hot smooth water, under a cloudless sky—

> " Idly as a painted ship
> Upon a painted ocean "—

with pitch oozing from her decks : a region of unbearable calm, broken occasionally by violent squalls, torrential rains, and fearful lightning and thunder "; (2) A tropical belt of high pressure rising irregularly ; and (3) A temperate and arctic region of generally low pressure, but with areas of temporary high pressure.

The bearing of atmospheric pressure upon the direction and force of the wind is now clearly established. From the daily returns at sixty stations in Western Europe the leading weather factors are communicated and are tabulated in the *Times* and other newspapers. Imaginary lines drawn through all the stations returning the same barometric pressure are called isobars, and it may from them be seen whether the pressure is normal or high or low. There are several of these lines upon the *Times* weather chart for the day, and it will be seen that they are in pressure a fifth of an inch apart. The varying distance asunder of these lines measures the varying force of the wind. The nearer together they are the steeper is said to be the gradient, a sharp gradient betokening a strong wind. For the direction of the wind there is a simple relation which is absolute. *Place your left hand towards the lower isobar, and your right hand towards the higher one, when the wind will strike your back.* This is Buys Ballot's fundamental

weather law, which is true not only for stormy periods
and the more marked atmospheric movements, but also
for the gentler movements of the atmosphere—a general
rule to which there is no exception.

Other winds of a periodic character are the land and
sea breezes experienced upon islands and coasts of the
warmer parts of the earth.   Their explanation is simple
enough.   They occur chiefly in tropical or sub-tropical
regions, because in those latitudes the night is never
short, the phenomena depending upon the difference   Land and
between the temperature of day and night, which is   sea breezes.
most marked in the tropics.   During the daytime the
land, as would be the case with any solid, becomes
hotter than the surrounding water ; and, when the sun
has well risen, say at nine or ten in the morning, an upward current of hot air is formed over the land.   This
is accompanied by an in-draught from the sea, which
constitutes the sea breeze, and which gathers force during
the day and in the early evening dies away.   At night
the land as readily and rapidly cools, whilst the sea
retains a higher temperature.   An ascending column
of hot air is developed at sea all round the coast,
and there is an out-rush of air from the land established to preserve the equilibrium, giving rise to the
land breeze which blows all night ; but it is not so
strong as the sea breeze, the water not being heated to
any very high temperature.   The distance from the
shore-line landward and seaward over which the movement is felt varies from a few miles to very long
distances.   The alternation of land and sea breezes is
strongly marked around the shores of the Mediterranean, and may faintly be perceived upon our own
coasts.   Vesuvius, when smoking, exhibits the effect   Illustra-
of this change, the smoke-cloud travelling out to sea at   tions of
night and being driven inland during the day.   Along   sea winds.
the Malabar coast of India this phenomenon is well
developed, the land breeze carrying far out to sea the
spice-laden breezes from the land, whilst, in the morning,

the inhabitants are refreshed with the cooler wind from the ocean.[1] Where large lakes occur some movement of the air of this kind has been recorded, the waters of a great lake being affected by change of temperature thirty-fold less than the surrounding land.   In some ports navigation is aided by this daily and nightly change, sailing vessels coming in with the sea breeze and leaving port with the land breeze at night.   It is always profitable for the untravelled to hear from a truthful eye-witness a description of natural phenomena, such as the following summary condensed from the account by Captain Dampier: "These sea breezes do commonly rise in the morning about nine o'clock, sometimes sooner, sometimes later; they first approach the shore so gently as if afraid to come near it, and ofttimes make some faint breathings, and as if not willing to offend they make a halt and seem ready to retire.   I have waited many a time, both ashore to receive the pleasure, and at sea to take the benefit of it.   It comes in a fine, small black curl upon the water, when all the sea between it and the shore not yet reached by it is as smooth and even as glass in comparison.   In half an hour's time after it has reached the shore it fans pretty briskly, and so increaseth gradually till twelve o'clock; then it is commonly strongest, and lasts till two or three a very brisk gale, and about five o'clock, sooner or later, according as the weather is, it is lulled asleep, and comes no more till the next morning. . . . When the sea breezes have performed their offices of the day, by breathing on their respective coasts, they in the evening do either withdraw from the coast, or lie down to rest; then the land winds, whose office is to breathe in the night, moved by the same order of divine impulse, do rouse out of their private recesses,

A sea-captain's account.

---

[1] The relative speed of cooling exhibited by solids as contrasted with water may be observed by noticing how quickly the temperature of a hot brick is reduced compared with the rate of cooling of hot water of equal volume and temperature.

and gently fan the air till the next morning ; and then
their task ends, and they leave the stage." This navigator further shows the importance of these winds to
Jamaica, and to the fishermen upon the coasts of Peru
and Mexico. In Jamaica the sea breeze, from its
health-bearing influences, is called "The Doctor," and
persons, when meeting each other in the morning, remark
upon its appearance, the conditions of the climate being
otherwise fixed. Where an island lies, like Jamaica,
under the full influence of the trade winds, the action is
modified, the land breeze being scarcely able to effect
the temporary reversal of the prevailing wind.

The monsoons[1] or season winds of India may be
regarded as land and sea breezes upon a large scale,
substituting for the day the summer half-year, and for
the night the winter half. The continent of Asia exhibits the greatest compact mass of land in the world.
When the sun goes northwards, India and South Central   The mon-
Asia become highly heated, and an ascending column of   soons.
hot air on a very large scale is produced, in rivalry of
the rise of heated air in the great equatorial chimney.
During the winter half-year the movement of wind
across India towards the equator is uninterrupted, the   Seasonal
current being deflected as explained under trade   winds.
winds, and this regular current is the north - east
monsoon, which, as proceeding from the north and over-
land, is dry and cool, rendering the winter half-year
the best time for Europeans to travel in India. During
the summer half-year, but beginning on various dates,
the sun in its northern declination heats up the land
of India and Southern Asia, producing an ascensional
current of immense strength. The northerly currents
playing upon India are not affected. The winds no
longer pass over the land towards the Indian Ocean,
but are carried upwards. There is now no passage of   Wet and
wind from India to the sea ; but, on the contrary, the   dry mon-
                                                          soons.

[1] From the Persian word *Mousum*, meaning a season, corrupted into
monsoon.

sea winds rush in to supply the vacancy caused by the masses of hot air ascending, and in this way the south-west or summer monsoon arises, the westerly element being due to the earth's rotation, as in the case of the return trades. The summer monsoon is hot and wet, the heavy rains of India being associated with this wind. The change from one monsoon to the opposite, known as the "breaking up of the monsoons," is accompanied by tremendous storms and unsettled weather for a month or more. Descriptions are not wanting of the sublime scene often presented by sea and land during these storms. An eye-witness, writing from Madras, speaks of the misty and reddened atmosphere; the rain falling in sheets; the roaring of the wind out-roaring the sound of the surf; its force bending the cocoa-nut palms almost to the earth; the lightning in broad sheets of flame, and the pealing thunder. "Fish upwards of three inches long were found upon the flat roofs of the houses in the town during the prevalence of the monsoon, either blown from the sea by the violence of the gale, or taken up in the water-spouts, which are very prevalent in this tempestuous season." These storms, however, clear the air, and after one of these visitations "the face of nature is changed as if by enchantment," the heats are subdued, vegetation is refreshed, the rivers are full, the air clean, and life becomes pleasant.

Storm-winds.

The winds being primarily dependent upon the sun's influence, the greatest storms are possible only in tropical or sub-tropical areas by sea or land. The law of storms has been fully established, and descriptions of tempests have been preserved; but no language can do justice to these occurrences which, with the exception of volcanic outbursts or earthquakes, are the most dramatic of the phenomena of nature. Only occasionally have we in England a distant approach to a really great tempest. In the storm of 27th December 1703 the power and sound of the wind were both appalling.

An English storm.

Over a hundred people were killed by the falling of houses, including the episcopal palace at Bath. Eight thousand people perished in the Thames, Severn, and round the English shores. In London thousands of chimneys fell; many thousands of sheep were destroyed, and the Eddystone Lighthouse was overthrown; for light as is the air when at rest, it acquires tremendously destructive effects when driving at a speed exceeding that of a racehorse.

The hurricanes of the West Indies, and the Indian Ocean, the Pacific, and the Chinese Seas are rotatory storms or cyclones whose behaviour is accurately known, and whose course can be predicted, so that vessels, by skilful navigation, may escape destruction. In the West Indies, as in the western part of the Indian Ocean, the general course of the storm is from the equator northwards and southwards, bending into a curve known to sailors as the "storm U." The wind travels in a progressive spiral circling in the southern hemisphere, always in a particular manner, viz. in the direction of the hands of a watch when placed face upwards. In the northern hemisphere the hurricanes originating in the West Indies circle backwards, moving in a direction contrary to that of the hands of a watch, or "backing," as sailors put it, still describing the "storm U" in their general direction. The velocity, the lifting power, the distance traversed, and the destruction caused on sea and land by these cyclonic storms may be better imagined than described. In the cyclone of 1830, originating near St. Thomas's Island in the West Indies, the storm route measured 1800 miles, and the storm travelled 25 miles an hour. The electric tension during one of these storms is often remarkable. On one occasion (in 1831), in the West Indies, part of a forest was destroyed without the trees being blown down, by, as it is supposed, the highly electric state of the atmosphere. In the hurricane of 1837, the scene in St. Thomas's Harbour was

The law of storms.

West Indian storms.

one of awful destruction.   A ship's log records thirty-six ships totally wrecked, a hundred sailors drowned, and the harbour a mass of wreckage.   Some of the houses were turned completely over ; and in the fortress six-pounder cannon were scattered about.   Great storms are reported pretty frequently from this and other hurricane regions.   There does not appear to be any appreciable diminution in the force of cyclonic storms, but vessels are now better equipped with steam power.   In the tempest which struck near the island of Samoa in the year 1890, the war vessels of several nationalities were overwhelmed.   An English ship, the *Calliope*, alone rode through the great gale uninjured.

Great hurricanes.

The vessel had 10,000 horse-power, and after a tremendous struggle, lasting two days, the engines, being urged to the utmost, slowly made headway against the wind, an act of seamanship that will long be remembered.   The winds thus occasionally defy the power of man and sometimes overthrow him; but ordinarily they are his servants.   A modern geographer has thus summarised their operations : [1]   "By their agency the moist and heated atmosphere of one region is transferred to another, and on their agency also depend, in a great measure, those currents of the ocean which are ever producing interchanges between the colder and warmer surface waters of different latitudes.   Besides the great climatological functions they are intimately concerned in the production of rains and other aqueous phenomena ; while their incessant commotions tend to preserve the atmosphere in ever healthful equilibrium.   They are the great bond between the land and water surfaces, transferring the moisture of the one to the thirsty uplands of the other, and the dry cold air of the former to disperse and rarefy the humid and depressing atmosphere of the latter.   Geologically, winds have considerable effect in removing, piling up, and re-assorting all loose super-

[1] Page's *Advanced Text Book of Physical Geography.*

ficial matters, as the sand dunes of the sea-shore, and
the sand-drift of the desert; while through the agency
of the ocean waves, which are created by their power,
important changes are produced along every shore of
the world.   In their gentler manifestations they assist
in the fertilisation of plants and in the dispersion of
their seeds; while in their fierce demonstrations their
track is marked by ruin and devastation.   By their
impulse the commerce of distant nations is wafted from
shore to shore, and, fickle and fitful as they may appear,
man not infrequently avails himself of their power to
turn the wheels and shafts of his machinery."

X

THE FORCE AND THE FILIGREE OF FROST

" 'Tis good to see
How multiform are lavish Nature's charms !
Discerning eyes have ever their delight,
And wonders countless greet us everywhere,
Alike in mountain path, in woodland glade,
In height majestic, in the lowly vale,
As in the matchless frost scenes deftly wrought
By broidering fingers o'er our window panes."
S. E. Tonkin.

BORN of a northern race and occupying a northern land, yet how seldom do we lie under the real power of frost. As near to the Polar realms as inhospitable Labrador and desolate Kamtchatka,[1] we yet realise but faintly the giant sway with which in some *England not a land of frost.* countries King Hrymir wields his sceptre, and English people have but few opportunities of seeing to perfection the delicate grace with which his frosty fingers transmute the waters and the vapours into "ornaments beyond the reach of art." Yet during the past twenty years we have occasionally experienced winters of almost Arctic severity, which may serve to exemplify something of the energy and the beauty with which

.

[1] Manchester, for example (53° 30' N.), is farther north than Irkutsk (52° 20' N.), the capital of East Siberia, which is reputed the coldest town in the world.

this great natural agent carries on its operations.[1]
Water turning to crystal enlarges to the extent of one-
eleventh of its bulk, so that it floats.   This expansion
is accompanied with extraordinary force.   If unen-
closed, no great effort is perceptible, just as unconfined
gunpowder in small quantities burns without much commotion; but no sooner is freezing water or exploding
gunpowder closely confined than prodigious are the

Fig. 83.—Expansive Force of Freezing Water.

results.   In the mine the blast rends the coal seam,
and amongst the mountains great disintegration and
ruin are wrought through the imprisonment of water
in the crevices of rocks exposed to nightly frost.
Bombshells filled with water have been burst when the
water was frozen (Fig. 83).   But a simpler experiment
may illustrate the point.   Take a small stout iron
bottle—it may be an inch thick—and furnish it with
a carefully-fitting screw plug, for an ordinary stopper

Expansive force of frost.

---

[1] At Torquay the thermometer, during the winter of 1860, registered
− 20° Fahr. ; at London, in 1867, − 3° Fahr. ; at Christmas, 1878, − 4°
Fahr.   The winters of 1891-93 were also unusually severe.

will not be nearly strong enough. Now fill with water which has been boiled, and tighten down the screw. Next place the bottle in a freezing mixture.[1] It will be as well to watch the action from a distance. Presently a sharp snap will announce the end of the struggle between the expanding water and the cohesion of the metal, and that frost is victor will be evidenced by the fractured bottle with its kernel of ice. This experiment, and the breakage of so many water pipes in a similar manner during very severe weather, may enable one more closely to follow the description of a traveller amongst the Himalayas, "the abode of snow," when he relates how the country exhibits several miles in the vertical direction of snow-covered mountains, and how stone avalanches, the accumulated frost-riven fragments, loosened by the foot of the traveller, are precipitated with deafening noise and irresistible momentum into the lower ground.

Crystal water has another property—one which has not been very long understood. Scientific men have been exercised as to how glaciers preserve their unity when following the windings of a valley, sometimes twisting at an abrupt angle, and yet appearing as an unbroken river of ice, save for the crevasses, which do not seriously interfere with its general integrity (Fig. 84). Several theories had been advanced and refuted, until experiment came to the rescue, and showed that ice has a remarkable property of refreezing, or regelation—to use the now accepted technicality—even under conditions that would seem impossible. The experiment may, in these days of manufactured ice, be easily made. Procure two pieces of ice, and flatten them upon a heated surface. Now press the flattened pieces together, when they will weld into one block. This will happen even if the room be quite warm, and may,

Glacier motion.

---

[1] An accessible freezing mixture is made by taking two parts by weight of snow or pounded ice, with one part of salt, which will reduce the temperature nearly to zero Fahr.

FIG. 84.— Glacier of Zermatt.

with care, be made to succeed in water as hot as the
Regelation. hand can bear.   Another experiment teaching the same
lesson may be arranged.   Obtain a small block of ice,
about as big as a brick.   Elevate it upon a footstool or
other support.   Pass over the middle of it a foot or
two of copper wire stretched tightly, and made to press
upon the ice by a flat-iron or other convenient weight
suspended at each side.   The wire will gradually cut
its way into the ice, and the severed parts will simul-
taneously refreeze, and secure the wire firmly in the

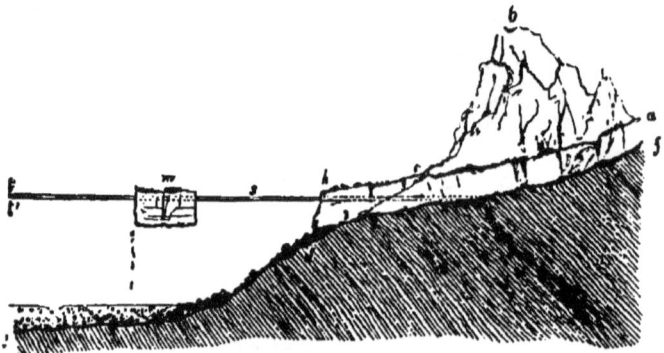

FIG. 85.—Formation of Icebergs.
    The glacier (*a*, *h*) descends from mountainous ground (*b*) to the sea-level (*s*),
    bearing moraine stuff on the surface, pushing on detritus below (*d*), and
    sending off icebergs (*m*), which may carry detritus and drop it over the sea-
    bottom ; *t*, *t'*, *g*, lines of high and low water.

block of ice in a curious manner.   If sufficient time be
allowed the wire will cut its way right through.   The
Ice can be teaching of this experiment is that ice, unyielding and
moulded.   rigid as it ordinarily is, will melt when pressure is
applied, and take any shape at will; but when the
pressure is removed regelation sets in.   This capability
of being moulded is so marked that an eminent physicist
—Professor Tyndall—speaks of the possibility of making
ropes and knots of ice.   There may be other physical
forces at work, but regelation plays an important part.
    Thus we are led to understand how the Alpine

glacier, laboriously grinding down a valley, encountering the stubborn rock, is melted at the opposing edges
by the great pressure from behind, and so breaks and
slowly flows round the obstacle like dough or treacle.
The unusual pressure being removed, it becomes solid
ice again, following the sharpest turns of the descent,
and everywhere presenting the appearance of a compact
river of ice.

There is evidence that the north of England was
once covered with such glacial masses. The rounded
tops of some of the hills of the Pennine range are
probably due to the continued passage of glaciers over
them when the country lay under a thick ice sheet, as
Greenland lies now, and the gravel pits and clay beds
of these parts are other memorials of the vanished
reign of ice.   But now, as no English mountain pierces
the snow-line, we must journey to other lands to
understand the true might of a glacier; or we may
read the descriptions of eye-witnesses.   In Switzerland
the glaciers dissolve as they reach the warmer lower
valleys, and immense burial mounds of rubbish carried
from the heights mark the place of their dissolution.
In northern lands, however, the ice river reaches the
sea (Fig. 85).   If the termination of the valley be low,
the nose of the glacier is projected into the water, by
whose buoyancy the cohesion of the ice is overcome,
and the broken fragments, floating away with the
currents, begin a new career as icebergs.   If, on the
contrary, the valley be high, the glacier overhangs the
cliffs until it can hold no longer, and masses fall away
with a mighty splash.   Often of huge size, they go
aground in deep water, and are amongst the well-known
terrors of Arctic navigation.   An explorer thus describes the change from the glacier to the iceberg:—

"Imagine a mighty river (as at Fig. 86), of as great
volume as the Thames, started down the side of a
mountain, bursting over every impediment, whirled
into a thousand eddies, tumbling and raging on from

The Ice Age of Britain.

Formation of icebergs.

Ice rivers.

ledge to ledge in quivering cataracts of foam, then suddenly struck rigid by a power so instantaneous in its action that even the froth and fleeting wreaths of spray have stiffened to the immutability of sculpture. Unless you have seen it, it would be impossible to conceive the strangeness of the contrast between the actual tranquillity of these silent rivers and the violent descending energy impressed upon their exterior. You must remember, too, that all this is upon a scale of such prodigious magnitude that when we succeeded subsequently in approaching the spot, where, with a leap like Niagara, one of these glaciers plunges down into the sea, the eye, no longer able to take in its fluviatile character, was content to rest in simple astonishment of what then appeared a lucent precipice of gray-green ice, rising to the height of several hundred feet above the masts of the vessel."

Unless we have seen it, the writer well observes, we cannot completely follow. Some faint realisation of Arctic conditions may be gathered from the experience of a very severe winter even in England, the country lying for weeks under a snow sheet, outdoor work given up, canal traffic stopped, trains immured in the snow-drifts, vehicular movement impeded or suspended, the gardener frozen out, and the farmer suffering heavy loss.

But severe winters have their relief. What an animated picture does the frozen surface of a lake present when covered with skaters; and at night, by means of the electric light, the recreation may still be pursued. Here is such a night scene. The water is artificial, but is some acres in extent, and the snow has rounded off the hard lines of the banks that would otherwise mar the artistic effect. There is no fear of accident; the water is but shallow, and none of the Humane Society's men are needed here. This is the eighth week of continued frost, and practice has made the skaters perfect. Many ladies are present, and they

Ice
exercise.

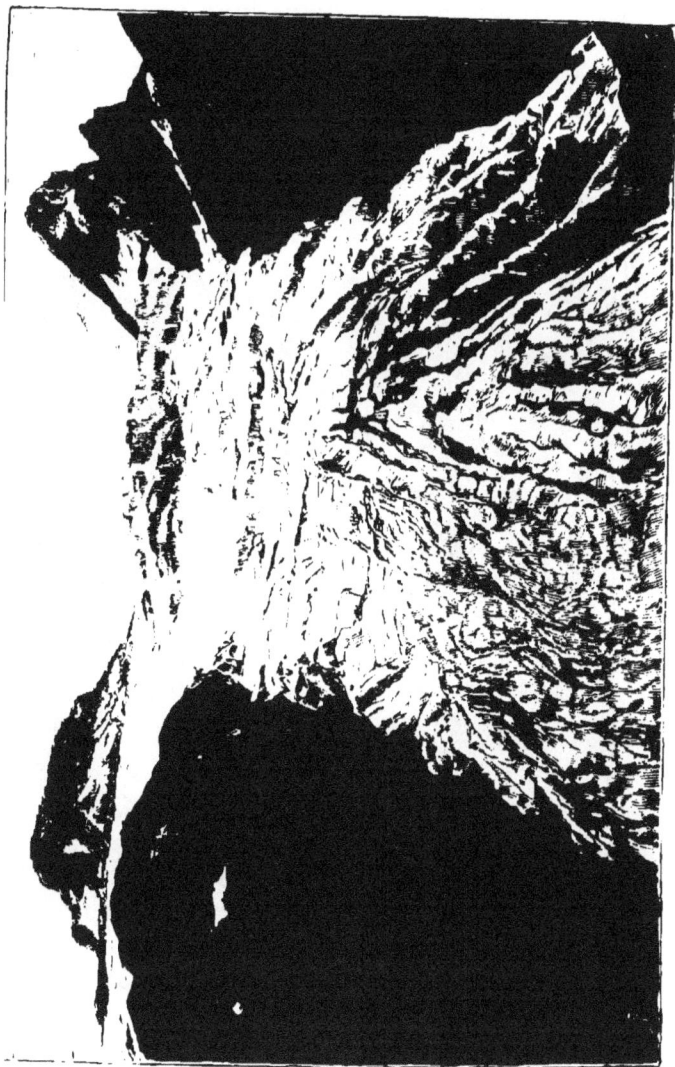

Fig. 86.—Glacier of the Blümlis Alp.

Ice exercise by the electric light. glide with more grace, if with less speed, than the male skaters. In the centre of the lake, upon an island, the blue-white electric beacon streams out steady and bright, almost deluding us into the belief that it is day. Little is to be heard beyond the ring of skate irons,

Fig. 87.—Forms of Snow Crystals (Scoresby).

and the deep responsive growl of the ice. If the English are "taking their pleasure sadly, after the manner of their nation," it is hearty enough if quiet ; and one after another the skaters come to land flushed and triumphant. Somewhat apart from the circling crowd are the famous skaters, executing the special figures of the icy art. Old Hrymir seems holding high festival with the genius of modern science as master of the ceremonies.

If snow and ice are beautiful in their broad effects Ice
they are no less so when the constituent particles flowers.
are separately examined (Fig. 87). Ice is crystal
water, and is supposed to be built up symmetrically
of stars, each having six rays, neither more nor less,
just the number of parts in the perianth of the

Fig. 88.—Ice Crystals.

lily. Generally hexagonal, the frost flowers exhibit
great diversity in detail. Captain Scoresby, in his
Arctic voyages, has figured no less than ninety forms;
but the pencil can but faintly render their beauty. It
is best to see these ice flowers thrown upon a screen
from a thin plate of ice by means of a good magic
lantern and the electric light. As the ice dissolves
under the warmth of the electric beam the crystals
blossom out all over the screen in the most enchanting

Molecular constitution of ice.

Ice and glass.

manner (Fig. 88). We can now the better see the force of the contrast which the physicist has drawn between the molecular constitution of ice and of glass, which, in externals, it so closely resembles. The glass particles will not so respond. They are in a state of strain—heterogeneous and harsh. There is in a piece of glass, according to Professor Tyndall, no more music than in a discordant Billingsgate cry, whilst ice in its minute structure may be compared with the finest harmony. Snow is equally attractive from this point of view. With a pocket lens, taking care not to breathe upon the specimen, their characteristic crystalline form may be seen. They grow to larger and more complicated forms of beauty under the keener frost of Alpine heights. A mountaineer, whose distant view is for a time cut off, may find occupation in inspecting the snow flowers under his feet. "Nature," says Professor Tyndall, "seemed determined to make us some compensation for the loss of all prospect, and thus showered down upon us those lovely blossoms of the frost; and had a spirit of the mountain inquired my choice—the view or the frozen flowers—I should have hesitated before giving up that exquisite vegetation of the frost flowers moving to music and ending by rendering that music concrete."

Frost flowers.

Decorative effects of frost.

Hoar frost is often finely decorative, as seen upon the window panes (Fig. 89). The most unpretending objects are touched into lovely effects; commonplace chiselling, the gate posts, and even the fences on the roadside are spiritualised by the delicate ornamentation of the frost. The outline of the shrubs and bare trees is defined by the crystal growth. Conspicuous are the hollies, every leaf margined with white, and, rivalling them in beauty, the conifers. These are good opportunities for studying tree form. The erect beech, the gnarled oak, and the stately elm, each is indicated, the characteristics of trunk and branch intensified rather than concealed by the frost foliage. But

FIG. 89.—Frost Flowers.

R

the palm is with the hawthorn hedges. In the thaw
they are brown and dull-looking enough, but now
Frost fili- every spray shows in pleasing curves, gleaming like
gree. banks of shining jewels, whether sparkling under the
low winter sun, or lighted up by a softer radiance
when comes forth—

> "That orbed maiden
> With white fire laden,
> Whom mortals call the moon."

The tracery upon the window pane is often exquisite,
especially when the moonlight, struggling through the

FIG. 90.—Arctic Iceberg seen on Parry's first voyage.

semi-transparency, lights up the surface with glittering
points of light, bringing to memory another scene of
moonlight and gems :—

> "There drew he forth the brand Excalibur,
> And o'er him, drawing it, the winter moon,
> Brightening the skirts of a long cloud, ran forth
> And sparkled keen with frost against the hilt :
> For all the haft twinkled with diamond sparks,
> Myriads of topaz-lights, and jacinth-work
> Of subtlest jewellery."

In Polar regions and on high mountains frost force is paramount. The Arctic and the Antarctic Oceans are covered with icebergs of gigantic dimensions (Fig. 90). For every foot above water of these ice mountains there are 7 or 8 feet below, depending upon its cargo and upon its shape, so that a berg 100 feet above the sea would have a total height of not much less than 1000 feet, or a fifth of a mile. Hence they are occasionally stranded even in deep water; and as they may weigh thousands of tons, the impact from even a slight motion is terrific. Captain Scoresby counted five hundred icebergs in sight at one time, and three

Fig. 91.—Tabular Iceberg detached from the great Antarctic Ice-barrier.

hundred have been passed by a steamer in the voyage to America from England. For such an army of icebergs there must be an extensive glacier-producing country; and, in Greenland alone, there are some 30,000 square miles of glacier-forming ground. Icebergs present every variety of appearance. There is generally a high side to windward and a lower sloping side to leeward. They ride firmly, the waves breaking against them as against a rock.[1] Such is their weight

[1] Lieutenant Parry measured an iceberg 4169 yards by 3869, and 50 feet high, which was aground in 60 fathoms water, and was computed to weigh 1,000,000,000 tons; another stood 120 feet out of the water.

244 SHORT STUDIES IN NATURE KNOWLEDGE x

that they crash through fields of ice. In contact with
Force of ice-fields or each other, and under the dissolving action
icebergs. of the warmer seas into which they are carried by
currents, they gradually break up, huge pieces (called
" calves ") rising upwards with tremendous force. In
consequence of such losses the stability is disturbed,

Fig. 92.—The *Vega* and *Lena* moored to an Ice-floe.
On the morning of the 12th August 1878. (After a drawing by O. Nordquist.)

and an iceberg may take up a new position, or may
even roll right over with immense commotion.

Their appearance has often been described. In
colour they resemble cliffs of chalk, or white or gray
marble, and, where fractured, are greenish-gray or
emerald-green. At night they have a faint effulgence,
and in a fog show dark. In a gale they are steady, on
account of their weight, and even seem to move to
windward on account of the surface ice flowing past
them. Their forms are often fantastic and beautiful.

To the eye of fancy they resemble floating towers,
castles, churches, obelisks, and pyramids, or may show
solidly in great tabular masses.   There is often a mist
enveloping them, and, in their neighbourhood, the   Polar ice.
thermometer sinks rapidly.   They are a menace to
Polar navigation, many a good ship having been lost in
collision with them.   They are occasionally of use, as,
in a storm, a vessel may shelter on the leeward side.
As agents of change they play an important part in
conveying fragments of rock from their parent mountain.
The areas where icebergs dissolve will one day be
found covered with these travelled stones, or erratic
blocks.

Ice - fields (Fig. 92), which are formed upon the
sea—not inland, like icebergs—are inferior in thickness, but of enormous superficial extent, and are,
in some respects, more wonderful, if not more impressive, than icebergs.   Some of them are 20 or 30
miles across, and may be 50 or even 100 miles
in length, and from 10 to 15 feet in thickness.
The weight of such a mass of ice may be easily expressed in figures, but it is almost impossible to realise
it ; and when such a moving mass comes into collision
with a vessel or another ice-field, the shock is appalling.
Sometimes these immense floating masses acquire a
rotatory movement, under which the outer edges have
a velocity of several miles an hour.   A field so moving
sometimes strikes another at rest or revolving in an
opposite direction.   " A body of more than ten thousand
millions of tons in weight, meeting with resistance
when in motion, produces consequences which it is
scarcely possible to conceive," says Captain Scoresby.   Ice-fields.
" The weaker field is crushed with an awful noise."
Sometimes the destruction is mutual ; pieces of huge
dimensions and weight are not unfrequently piled upon
the top, to the height of 20 or 30 feet, while a proportionate quantity is depressed beneath.   The ocean swell
fortunately generally disintegrates the ice-fields into ice-

floes, with which the navigator may more easily contend. The noise accompanying these movements is described as resembling that of complicated machinery or distant thunder. Besides the iceberg, the ice-field, and the floe (which is a field on a small scale, say of half a mile or a mile in diameter), sailors speak of drift ice, which is formed of smaller irregularly-shaped pieces; of brash ice, which is still more divided; and of bay ice, which is newly formed on the sea. A hummock is a mass raised above the level of the ice. Pack ice is drift ice of large extent, open or close. Open ice or sailing ice is where the pieces are separate, so as to admit of vessels sailing; and a "lane" is where there is a narrow opening in the midst of the ice. The South Pole is far more encumbered with ice than the North (Figs. 93 and 94).

Forms of Polar ice.

When sea water freezes most of the salt is deposited, except that in the pores of the ice itself, so that sea ice when thawed gives fresh water. Salt-water ice is blackish in the water and white or gray in air. Fresh-water ice is fragile and so hard that the edges cut sharply, and Captain Scoresby points out that it is transparent enough to form a burning glass, "that with a lump of ice, of by no means regular convexity, I have frequently burned wood, fired gunpowder, melted lead, and lit the sailors' pipes, to their great astonishment."

Physics of ice.

The force of frost is also well illustrated by the glaciers of Polar countries or elevated regions. In its origin a glacier is snow only, formed above the snow-line or "the line along which the quantity of snow which falls annually is melted, and no more." Below this line each year's snow is completely cleared away by the summer heat. Above it a residual layer abides, which gradually augments in thickness from the snow-line upwards. These residual layers still travel downwards. At first the snow is dry and intensely cold; but descending it is warmed until it reaches melting

NORTH POLAR REGIONS

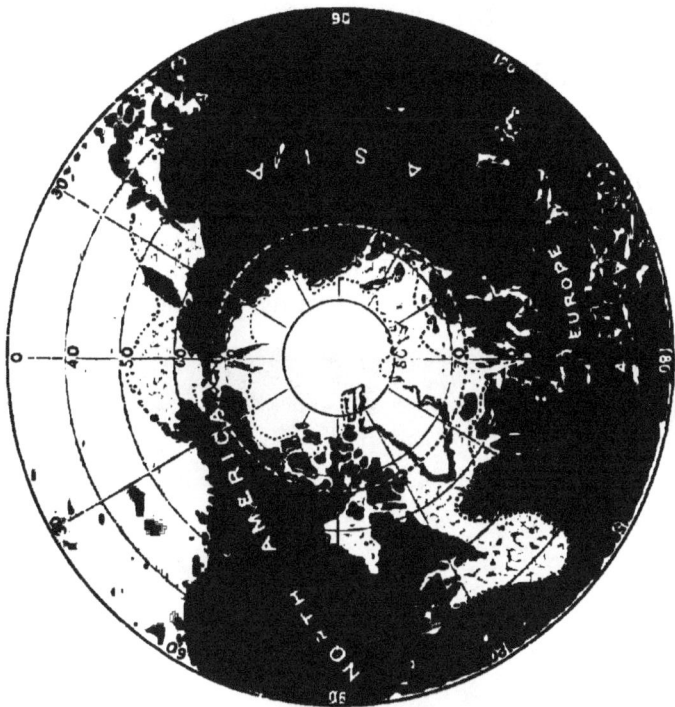

SOUTH POLAR REGIONS

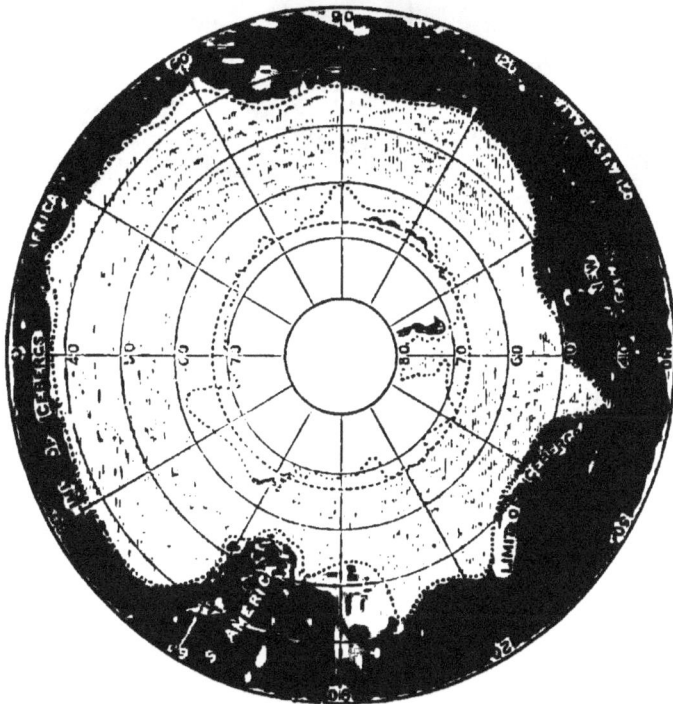

Fig. 93.—Diagram of the approximate extent of Permanent and Floating Ice around the North and the South Poles. (After Petermann.)

point, and there is at length formed a mixture of snow
and frozen water known as the névé, while the portion
which is clearly below the snow-line is the glacier
proper, and the texture is ultimately that of quite transparent ice.   Glaciers are all in steady motion, but
the rate is so slow that it is imperceptible to the senses,
and only to be established by experiment.   The Mer
de Glace travels about 100 yards per annum, but the

FIG. 94.—The Antarctic Ice-wall and Icebergs.

slope of the valley has an important effect upon the
rate.   A glacier has all the aspect of a river.   How it
turns sharp corners in its descent has been already
explained.   But the surface is full of chasms or *crevasses*,
which interfere with locomotion.   Along the sides of the
glacier are walls 50 or 60 feet high, formed of fragments
of rock known as lateral moraines.   When the glaciers
of two valleys unite, two of the side streams of stones
join, in the centre, to form a *medial moraine* (Figs. 95

and 96). Ultimately, on reaching the lowest point of the valley, several lines of rock may be found. The larger stones upon the glacier intercept the sun's heat and prevent the ice from melting immediately underneath, and in this way glacial tables are formed. Fragments of rock falling into the crevasses reach the bottom of the glacier, and score the fixed rocks beneath, the scratches or striæ remaining upon the rocks of the

Fig. 95.—Glacier with Medial and Lateral Moraines.

valley long after the glacier, or the glacial period, has completely passed away. Many of the valleys of Northern England are so scratched, telling of the time when this country lay under a huge glacial sheet, like Labrador and Greenland at the present time. The ice, when free from stony fragments, grinds heavily over the rocks and polishes them smooth. Just as water wears away stones, so the crystal water of the glacier wears down the rocks, the abraded material colouring the waters which flow under the glacier; and the stream, charged with this fine matter, settles down in the lower ground into pools, where it deposits the sediment which forms clay beds of the future. Glacial phenomena have been well observed and reported upon in detail by several eminent men of science.

Movement of glaciers.
· It has been demonstrated that glaciers move steadily and, for a given period, uniformly ; that the speed is slower at night, and in the coldest weather ; that the speed varies with the inclination of the valley ; that the centre moves faster than the sides, and the surface

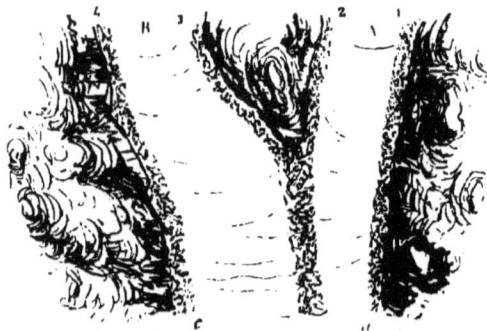

Fig. 96.—The union of two Glaciers, showing junction of two Lateral into one Medial Moraine.

more rapidly than the parts below. The "dirt bands" are brownish bands bending down the glacier, and having a wave-like form, which is produced by ridges on the ice surface being filled with the finer waste of the mountains. Glaciers not unfrequently exhibit fine caverns with icicles hanging from their roofs, and with walls of a beautiful green or blue colour. The actual examination of one of these glaciers in all its parts, effects, and phenomena, is very interesting, so that mountaineering in the Alps and amongst glaciers is steadily increasing in popularity, the tendency now being to go farther afield and to attempt the loftier heights of the Himalayas, the Andes, and the Caucasus.

FIG. 97.—The Mer de Glace.

In the Alps some of the glaciers are miles in length. The Mer de Glace is 7½ miles long (Fig. 97). In the Himalayas they are of yet larger dimensions, corresponding with the proportions of the range. In New Zealand there are some fine examples; and in Greenland there is a glacier 60 miles in length.

Where glaciers dissolve before reaching the sea the waters flowing from them form rivers. The Rhine, Rhone, Po, Garonne, Ganges, and Indus are of this character.

Glacial work.

The effects in nature of glaciers are various and important. They carry down the waste of the mountains into the lower grounds; they hollow out the rocks where the pressure is greatest, and form lake basins, generally of small size; but some of the larger lakes, *e.g.* Geneva, are partly, if not wholly due to the agency of glaciers. The boulder clays of England north of the Thames are clearly of glacial origin. No other physical agent could have placed the boulders which they contain in the positions which they occupy.

There are other proofs than those named above of a former Ice Age in this country—when man was only beginning his rule and the frost forces were supreme. The erratic blocks found throughout North Britain are stones foreign to the formation of the district, and were carried by icebergs or glaciers. In Scotland they were long attributed to the agency of wizards, warlocks, brownies, fairies, and the Evil Spirit, who were regarded by the Highlanders as following their favourite

Frost work in the past.

game of "putting" the stone; but science has dispelled these fancies and traditions, and these "foundlings" or foreign stones, by whatever name known, are clearly due to the great worker, frost. The presence of Arctic shells in the clays of England, as in the south-east of the country, is another indication of the Polar climate which once marked this land, for there is no surer guide to the character of a climate than the evidence afforded by the fauna and flora of the country.

The desire to penetrate farther the Polar regions, and to enter the fortresses of the powers of frost and snow, has long been a fascination.   The story of Arctic and

FIG. 98.—The Palæocrystic Sea.
From the "English Illustrated Magazine."

Antarctic exploration is one that cannot be told here; but it is sufficient to say that, notwithstanding tremendous obstacles—the intense cold, which causes the circulation of the blood to be unduly increased, and other very unpleasant physical sensations; the opposition offered to navigation by the fields of ice, and to sledge-travelling by the uneven nature of the icy surface; the absence of accommodation and the difficulty of carrying provisions—the margins, at any rate, of the Polar seas have been broken through. Captain Nares, in his expedition, carried his vessel and hoisted the English ensign in latitude 82° 24′ north, a higher point than any vessel had ever before attained. Captain Markham, by his sledge journeys, pushed northern discovery still farther, reaching 82° 48′, and sighted land lying in latitude 83° 7′. There was now an interval of but 400 miles separating this point from the North Pole. The difficulty of much farther progress by the means now used is shown by the circumstance that, although Captain Markham penetrated only 73 miles from the ship in a direct line, yet, owing to the difficulties of the route, 276 miles were travelled on the outward journey and 245 on the return. Some of the icy masses rose high as buildings (Fig. 98), and had to be passed over or a detour made, involving great loss of time. The cold was so great that with leather clothing of the warmest kind, sometimes weighing nearly half a hundredweight, the men were almost frozen as they walked. The meals had often to be taken in the open. Hot tea was found more beneficial than rum or spirits of any kind; but the ice had to be thawed to form water for the supply, and the food had to be eaten in a half-frozen state.[1] Captain Nares, in his official report, states that upon the *Alert* exploring vessel a temperature of 73° below zero was registered, and upon the sister ship, the *Discovery*, 70° below zero. Quicksilver was frozen for fifteen successive days, and the effects of such low temperatures were remarkable in the extreme. Brandy and spirits became solid; iron lost its nature and became brittle; and even the movement

Marginal notes: Captain Markham's sledge journey towards the Pole. Difficulties and encouragements of Arctic travel.

[1] The writer heard this journey described by Captain Markham, who mentioned that the sailors had to thaw the bacon with which they were supplied in their hot tea in order to eat it.

of the compass needle was rendered sluggish. There are, however, some advantages even in these low temperatures. The health of the crews was extremely good, the dry, if cold, air being very exhilarating; and it was only when lower latitudes were reached that colds and rheumatic affections were noticed. In such cold climates also there need be no anxiety as to the preservation of food, which will keep for weeks or even months, and be improved rather than depreciated by being frozen. The details of these journeys are worthy of being studied at length in the great books of Arctic travel, amongst which may be named those of Commander Scoresby. Arctic exploration is still in progress. England, America, Norway, and Sweden are emulating each other in this research, and, no doubt, new light will further break in upon us, and the remarkable operations of frost, whether in force or filigree, be more abundantly revealed.

FIG. 99.—Frost-work. Luna Island, Niagara.

THE EARTH'S FIRES

"He looketh on the earth, and it trembleth : He toucheth the hills, and they smoke."

"And, lo, there was a great earthquake, and the sun became black as sackcloth of hair, and the moon became as blood, and every mountain and island were moved out of their place, and every island fled away, and the mountains were not found."

WHEN the earth was young, and, according to the notions of the early inhabitants, every force in nature had its presiding divinity, prominence was given to the fire gods, especially in the mythology of South Europe, the Mediterranean being the earliest studied of the great volcanic areas. The pagan Greeks and Romans had a lively faith in the might of Typhon the hundred-headed, who, with fire-flashing eyes, was imprisoned beneath the heavy load of Etna, and of Vulcan, and the Cyclops, the forge-men of Jove, toiling in their fiery workshop. Through the Middle Ages but little progress was made in earth knowledge, although certain cosmic phenomena were carefully observed and accurately recorded ; and it is only within the past half century that crude ideas as to the nature of igneous phenomena have been displaced by scientific information. The distance from the surface to the centre of the earth — about 4000 miles—does not exceed a journey across the Atlantic Ocean ; yet no one has penetrated more than one mile below the surface, much

Gods of fire of the ancients.

less attempted to reach the centre. Information of the earth's interior state, proceeding by observation and mathematical reasoning, has been obtained with difficulty. The influence of early errors still lingers in our everyday speech; we still talk of the "crust of the earth" as in the early part of the present century, when it was generally believed "that this globe we live on is a stupendous but very thin bombshell charged with liquid fire." If the term "crust" be used, it is not in the same sense as formerly.

An important step onwards was taken when Cavendish determined the high specific gravity of the earth, and showed that this world, whilst of enormous dimensions, was also composed of very heavy materials. The earth was conclusively proved by these and subsequent experiments to be five and a half times heavier than water—that is, of greater weight than if composed of the heaviest stone.[1] Another physicist[2] has shown that it is not only exceedingly heavy, but that it is extremely improbable that any crust thinner than 2000 or 2500 miles could maintain its figure with sufficient rigidity against the tide-generating forces of the sun and moon to allow the phenomena of ocean tides and the earth's complicated motion to be as they are now.

No central sea of fire.

The globe heavier than solid stone.

The notion of a central sea of fire, or even of lakes of fire of any great dimensions, was gradually abandoned after the laborious researches of Mallet had demonstrated that the depth from which earthquake action was set up was very slight; that the average distance of the centres of explosion was about 15 miles below the surface, and that the greatest observed depth of an earthquake focus was 30 miles. In proportion to the diameter of the earth—8000 miles—this is inconsiderable, justifying the modern view that earthquakes are superficial movements, and not in their origin deep-seated, as formerly supposed.

The earth a solid.

Earth-quakes confined to the surface.

[1] Say twice as heavy as granite.
[2] Sir William Thomson, F.R.S., now Lord Kelvin.

Humboldt, travelling in the volcanic regions of South America, and in other parts of the world, had abundant opportunity of observing earthquake phenomena, and endeavoured to trace a relation between the state of the atmosphere—particularly as regards its temperature and barometric pressure—and the occurrence of earthquakes and volcanic outbursts. That there is some such connection has been felt by the Neapolitans, who say, "When Vesuvius grumbles bad weather is at hand." And several important earthquakes have been coincident with abnormal atmospheric conditions, the barometer, for example, being lower over the portion of Greece during the earthquake at Ischia than in any other part of Europe; and "earthquake weather" is spoken of in certain volcanic countries. This is a field of inquiry yet to be worked out. If earthquakes are not deep-seated in their origin, it is intelligible to suppose that the ruling force is external rather than internal, and it may be proved that the moon and the nearer planets, as their distances and attractive power vary, modify, if they do not originate, igneous action within the earth's crust; and that the sun, the dispenser of light, heat, and the gravitation centre of the planetary system, may be found to be influential in causing the fluctuations of forces within the earth's circumference that lead to the earthquake throb, the volcanic eruption, and the slower elevation and depression which characterise certain regions of the earth. The play of electric and magnetic forces may, in a minor and intermediate manner, have something to do with the case.

The first work is, however, to ascertain what are the undoubted facts.

There are three important volcanic regions associated with the great continents, besides several minor areas, where the surface of the earth is close to the igneous forces below—where volcanoes are in almost continual eruption, and earthquake shocks common.

Igneous phenomena probably governed from outside the earth.

The sun's control of the earth.

The great volcanic bands.

The European region is associated with the south of the continent — with the Mediterranean countries. It may be thought of as extending from the Caspian Sea to the Azores, from east to west over 1100 miles, and from north to south some 800 miles. The Black Sea, the Alps, the Cevennes, the region onwards to the Pyrenees, and, under the sea to the

Fig. 100.—Sketch of submarine volcanic eruption (Sabrina Island) off St. Michael's, June 1811.

Azores, 25° west of Greenwich in the North Atlantic, where a new island was thrown up in the year 1811 (Fig. 100), may be regarded as the northern limit. The southern boundaries of the Eurasian system stretch from the country of the Tigris and Euphrates, through Palestine, Arabia, and North Africa to the Canary Islands. In this area occur all the leading symptoms of igneous activity—volcanoes, hot springs, fumeroles,

oil springs, and occasionally earthquakes of extreme
violence.   This region, extending over the earliest
civilised countries, has been longest under observation,
and may be taken as the key to volcanic phenomena all
over the earth, in general character if not in intensity
of action.   Between the Caspian and the Black Sea are
the "Fields of Fire," where the earth has numerous
fissures from which gases burning with a blue flame
continually ascend.   This constant exhibition of the
fire forces must have had effect upon the native people,
who became fire-worshippers and erected shrines for
the fire divinities.   Here also are almost inexhaustible
springs of naphtha, which occasionally take fire in the
wells, or burn on the surface of the Caspian Sea with
remarkable effect.   Earthquakes periodically convulse
this district, as when, in 1840, Mount Ararat was shaken,
and torrents of melted snow with masses of earth
poured down its flanks, destroying the inhabitants
around, and laying waste the cultivated land.   The floor
of the Sea of Azof was, in the year 1814, partly broken
up, and, after a tremendous conflict between fire and
water, a new island was established.

In Palestine, and the lands spoken of in the Bible,
earthquakes were regarded as physical agencies by
means of which the Divine displeasure was signified.
To the dwellers in the Holy Land the earth was by no
means the *terra firma* which it is in Western Europe,
references to its instability being scattered through the
sacred pages.   "The Lord uttered his voice, the earth
melted"; "He made the earth to tremble," says the
Psalmist; and in the prophetical books are also
similar references: "Yea, ye shall flee, like as ye fled
from before the earthquake in the days of Uzziah, king
of Judah, and the Lord my God shall come, and I will
send a fire upon Teman, which shall devour the palaces
of Bozrah."   In the days of Herod the Great, Josephus
states that there was an earthquake; and in Syria
in the time of Tiberius, A.D. 19, it is computed that

The European volcanic regions.

The European igneous region.

Bible references.

FIG. 101.—Stromboli, viewed from the north-west, April 1874.

120,000 persons were destroyed. Proceeding westward
The Mediterranean. we come to Greece and Italy, where the fire forces
have from time to time gathered in their intensity,
whether to produce earthquakes or volcanic outbursts.
Stromboli, near the Italian coast, is almost continually
active (Fig. 101). The Isles of Greece, with Santorini
as a centre, have been the scene of repeated displays
of these, "the most dramatic of all the phenomena
of nature." Later manifestations occurred in the
Ionian Islands, in 1820, when a new island was
thrown up; in 1841, when an island near Zante dis-
appeared; and there were hundreds of destructive
shocks in 1894. In the Italian peninsula the destructive action has been still more remarkable. An
earthquake, of which the island of Ischia, the beautiful
sanatorium of the Neapolitans, was the centre, happened
in the year 1883. This island is repeatedly referred to
by the classical writers as a focus of fiery activity.
Early observations. The terrible incidents seem much the same, whether
related in the narrative of Strabo, the verse of Virgil, or
communicated in the staccato of the modern telegram.
In these regions they mark chronology, the people
reckoning from one fiery outbreak to another, as we of
more favoured lands reckon from one king's reign to
another. Close to Ischia is Vesuvius, the best-known
of all volcanoes. It has been studied for centuries;
one of the most complete records, dating from the
Vesuvius a typical volcano. beginning of the Christian era. Vesuvius may, therefore, be taken as typical, and the character and
sequence of phenomena exhibited by this volcano as
illustrative of volcanic action in greater centres far
away.

Vesuvius is of the conical form common to volcanoes,
and is close to the Bay of Naples, as seen in the illustration (Fig. 102). It rises to the height of nearly a mile
above the waters, and at its base is 10 miles in diameter
and 30 miles in circumference—quite the dominating
object of the country. The mountain is a mass of

erupted matter—of cooled lava streams piled upon
deep beds of ashes, and of ashes upon solid rivers of
cooled lava—a veritable cinder-heap of nature. The
form, especially near the cones, of which there are
many (eleven new ones being formed in the year 1861),
is being continually altered as the mountain is shaken
by repeated eruptions (Fig. 103). One of the earliest
recorded eruptions was in the year 79 A.D., and is

Fig. 102.—Vesuvius.

noteworthy as that in which the elder Pliny lost
his life. The incidents have been related by the
younger Pliny in an admirably clear narrative,
which, closely followed, will do something to bring
before dwellers in non-volcanic lands the terrors of
these times. All South Italy was shaken by this
eruption, and the cities of Herculaneum and Pompeii,
which stood upon the flanks of the mountain, were
destroyed. All round the shores of the Bay of Naples

First recorded eruption.

Pliny's account.

were the houses of the Roman patricians; and at the time of this outburst Pliny and his uncle were with the fleet near Misenium, situate upon the northern promontory of the bay. "On the 23rd of August, about one o'clock in the afternoon, my mother desired him (his uncle) to observe a cloud which appeared of a very unusual size and shape. . . . He immediately

An eruption of Vesuvius.

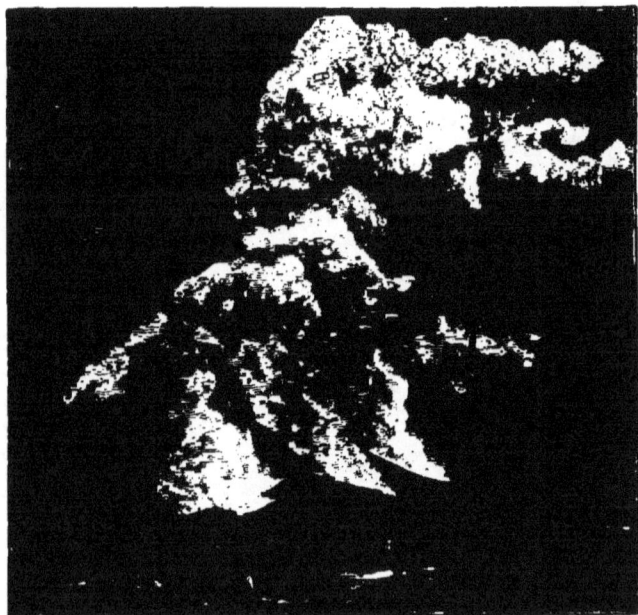

Fig. 103.—View of Vesuvius as seen from Naples during the eruption of 1872.

arose and went out upon an eminence, from which he might more distinctly view this very uncommon appearance. It was not at that distance discernible from what mountain this cloud issued, but it was found afterwards to ascend from Mount Vesuvius. I cannot give you a more exact description of its figure than by resembling it to that of a pine tree, for it shot up a great height in the form of a trunk, which extended itself at the top into a sort of branches, occasioned, I

First stages.

The pine tree of smoke.

imagine, either by a sudden gust of air that impelled
it, the force of which decreased as it advanced upwards,
or the cloud itself, being pressed back again by its own
weight, expanded in this manner. It appeared sometimes bright and sometimes dark, and spotted as it was
more or less impregnated with earth and cinders. This
extraordinary phenomenon excited my uncle's philosophical curiosity to take a near view of it. He ordered
a light vessel to be got ready, and gave me liberty, if
I thought proper, to attend him. I rather chose to continue my studies, for, as it happened, he had given me
an employment of that kind. As he was coming out
of the house he received a note from Rectina, the wife
of Bassas, who was in the utmost alarm at the imminent
danger which threatened her, for her villa being situated at the foot of Mount Vesuvius, there was no
way to escape but by sea. She earnestly entreated
him, therefore, to come to her assistance. He accord-
ingly changed his first design, and what he began with
a philosophical, he pursued with a heroical turn of
mind. He ordered the galleys to put to sea, and went
himself on board, with an intention of assisting not
only Rectina, but several others, for the villas stand
extremely thick upon the beautiful coast. When
hastening to the place from which others fled with
the utmost terror, he steered his direct course to the
point of danger, and with so much calmness and presence of mind as to be able to make and dictate his
observations upon the motion and figure of that dreadful scene. He was now so near the mountain that the
cinders, which grew thicker and hotter the nearer he
approached, fell into the ships, together with pumice
stones and black pieces of burning rock; they were
likewise in danger not only of being aground by the
sudden retreat of the sea, but also from the vast
fragments which rolled down from the mountain, and
obstructed all the shore. Here he stopped to consider
whether he should return back again, as the pilot

The eruption increases.

advised him. "Fortune favours the brave," said he;
"carry me to Pomponianus. . . .

"In the meanwhile the eruption from Mount Vesuvius
flamed out in several places with much violence, which
the darkness of the night contributed to render still
more visible and dreadful. But my uncle, in order to
soothe the apprehensions of his friend, assured him it
was only the burning of the villages which the country
people had abandoned to the flames. After this he
retired to rest, and it is most certain he was so little
discomposed as to fall into a deep sleep. . . . The
court which led to his apartment being now almost
filled with stones and ashes, if he had continued there
any time longer it would have been impossible for
him to have made his way out; it was therefore
thought proper to awaken him. He got up and went
to Pomponianus, and the rest of his company, who
were not unconcerned enough to think of going to bed.

Distress and uncertainty. They consulted together whether it would be most
prudent to trust to the houses, which now shook from
side to side with frequent and violent concussion, or fly
to the open fields, where the calcined stones and cinders,
though light indeed, yet fell in large showers, and
threatened destruction. In this distress they resolved
in favour of the fields, as the less dangerous situation
of the two, a resolution which, while the rest of the
company were hurried into by their fears, my uncle
embraced upon cool and deliberate consideration.

Flying from destruction. "They went out then, having pillows tied upon their
heads with napkins, and this was their whole defence
against the storm of stones that fell around them.
Though it was now day everywhere else, with them it
was darker than the most obscure night, excepting
only what light proceeded from the fire and flames.
They thought proper to go down farther upon the
The agitation of the sea. shore, to observe if they might safely put out to sea;
but they found the waves still running extremely high
and boisterous. There my uncle, having drunk a

FIG. 104.—Lava Stream.

draught or two of cold water, threw himself down upon a cloth which was spread for him, when immediately the flames and strong smell of sulphur, which was the forerunner of them, dispersed the rest of the company, and obliged him to rise. He raised himself up with the assistance of two of his servants, and instantly fell down dead, suffocated, as I conjecture, by some gross and noxious vapour, having always had weak lungs, and frequently subjected to a difficulty of breathing. As soon as it was light again, which was not till the third day after this melancholy accident, his body was found entire, and without any marks of violence upon it, exactly in the same posture that he fell, and looking more like a man asleep than dead."

The writer, with his aged mother, escaped with the people into the open. Their chariots pitched backwards and forwards, though drawn out on level ground, and blocked with large stones; the sea seemed to roll back upon itself, and to be driven upon its banks by the convulsive motion of the earth, and many sea animals were left upon the shore, from which the water had receded.   As the terrified crowd made their escape a thick sulphureous smoke rolled over them, and utter darkness overspread them.   Nothing was then heard but "the shrieks of women, the screams of children, and the cries of men ; some calling for their husbands, and only distinguishing each other by their voices ; one lamenting his own fate, another that of his family ; some wishing to die from the very fear of dying ; some lifting up their hands to the gods ; but the greater number imagining that the last day was come, which was to destroy both the gods and the world together. . . . At length this dreadful darkness was dissipated by degrees, like a cloud of smoke; the real day returned, and the sun appeared, though very faintly, and as when an eclipse is coming on, and every object seemed changed, being covered over with white ashes as with a deep snow."

This description has not been excelled as dealing with the superficial appearances, but other observers, including several Englishmen, have studied the behaviour of this mountain closely, the late Professor Phillips devoting a whole book to the subject, and as the action of one volcano fully understood becomes the key to all the rest, it may be helpful here to describe the character and sequence of the phenomena, with illustrations of exceptional energy, taken from the more remarkable eruptions. *Other eruptions of Vesuvius.*

In the year 1500 A.D., passing over numerous minor events, the flow of the river of fire we call lava was so great as to threaten very wide destruction. In this emergency the people sought to arrest the flow of the lava by carrying relics of the saints and sacred vestments in solemn procession. In 1631 A.D. there was another tremendous convulsion. The cone of the mountain was blown off, and it seemed as if the volcano would be moved from its seat. Villages around were ruined; vineyards and gardens were laid waste, layers of fine ashes being spread over the land for hundreds of miles, which tempests of rain converted into streams of mud. All this was to the accompaniment of horrible subterranean noises, which have been vaguely compared to thunder, or the rush of flights of rockets, whilst streams of lava (one of them issuing from a point over half a mile high) started down the slopes of the mountain and flowed for miles. The "pine tree" of smoke and vapour assumed an alarming aspect, and the sea oscillating with the land caused the waves violently to advance and retire. For six months these distressing scenes were renewed, and before the mountain came to rest 18,000 people were destroyed. In the seventeenth century there were seven, and in the eighteenth century thirteen eruptions. In that of 1779 an apparently blazing column rose two miles above the mountain's summit, illuminating the country around at the distance of twelve miles, giving light enough for *Great eruptions of Vesuvius.*

books to be read at midnight. The clouds of steam
and ashes were phenomenal, and volcanic bombs (as the
burning, flying rocks are called when rounded in their
passage through the air), some of a hundred pounds'
weight, battered down villages six miles away from the
mountain. In the present century there have been
extremely violent eruptions, and with the progress of

FIG. 105.—View of portion of a Lava Stream on Vesuvius (Abich).

physical science the volcano has been submitted to
close scrutiny of geologists assisted by instruments
of precision. In the year 1867 an observatory was
established upon Vesuvius, and a scientific party
travelled from England to Italy to watch an impend-
ing eruption, every incident and every phase being
carefully recorded.

The following is the sequence of events. Premonitory symptoms arc the failing of the wells and springs in consequence of the dislocation of nature's underground water pipes by the movement of the strata, the breakage accompanied by subterranean sounds. The cloud hovering over the mountain is greatly increased, sometimes to three or four times the height of the mountain. Earthquakes increasing in intensity are felt, and, if the

Fig. 106.—View of houses surrounded and partly demolished by the Lava of Vesuvius, 1872.

cloud be closely watched, lightning shafts (the " ferilli " of the Italians) may be seen darting through it, indicative of electric action. Next clouds of steam ascending are condensed into pouring rain. The puffs of steam increase and become rhythmic, and dust clouds are carried upwards to a tremendous height, falling upon Naples, or carried onwards to Rome or Athens, or even across the sea to Africa. The rain of stony matter also

increases, fragments flying to the height of half a mile
above the mountain top. All this time the lava is
rising like molten iron in a furnace, and, with a supreme
The climax effort, is poured forth in fiery streams, which at night
of an show like the fingers of a huge red hand. The speed
eruption. of the lava stream is at first considerable—several
miles an hour — and has been compared to the
movement of an English river, as the Severn, in its
middle course, losing pace as it gradually cools (Figs.
104, 105, and 106).

This marks the climax of the eruption, and is
succeeded by a quieter phase, in which various gases are
disengaged, which escape into the air or are condensed
around the throat of the crater. The steam still puls-
ates in regular jets, which, however, become fewer and
feebler. The more prominent gases noticed are hydrochloric acid, sulphurous acid, sulphuretted hydrogen,
nitrogen, carbonic acid, hydrogen, and carburetted
hydrogen. Metallic compounds are also deposited, as
iron pyrites, lead sulphide, zinc sulphide, together with
salt and sulphur. Valuable minerals and precious
stones, as garnets and malachite, are also collected, along
with the commoner gypsum, or sulphate of lime, and
pumice stone. The lava, which is found in quantities
sufficient to form a new surface for the earth, is a
composite material, its chief components being silica,
lime, alumina, iron, and magnesia, with traces of other
materials. A volcano is, in fact, an extensive mineral
Varied factory, and the geologists say that it would be easier
products to enumerate what is not discharged than what is.
of an Several of the products, as sulphur and pumice stone,
eruption. are of commercial importance, forming a source of
wealth in volcanic regions; but the popular idea that
little else than sulphur and lava are discharged is
erroneous. The lava stream does not, as is commonly
supposed, break from the crater like an overflowing cup,
but finds a weak place in the flank of the mountain
through which it can pass long before reaching the

summit. The notion that a volcano emits flames is also a mistake, the brilliant light alluded to being due to the abundant incandescent matter (as the white-hot carbon of an electric lamp), or the reflection upon the clouds of dust and steam of the intensely-heated interior of the crater.

Another feature of a volcanic eruption often overlooked is the mud lava—*lava d'acqua*—which is as destructive to property, if not to life, as the fire lava itself which so much impresses the imagination. Pompeii was overwhelmed, according to Pliny's description, by showers of powdered ashes, and by streams of mud and hot water. Most of the inhabitants escaped, but the flow of mud enveloped the city and the surrounding country, forming over the district a loose kind of stone which, in modern times, has been excavated, and the lost city restored, the mud lava preserving everything intact, even to the inscriptions and sign-boards of the first century.

The European igneous band has also been marked by earthquakes of no ordinary intensity. That of Lisbon is the best known. It occurred 1st November Earth-1755, and has been excelled in intensity only in quake of America and Asia. With a clear sky a sound of Lisbon. thunder was heard about nine in the morning; then without further warning the ground under the city broke up, and the falling houses and buildings crushed to death in about three minutes some 30,000 of the inhabitants. At this hour many of the churches were filled with people, for this was All Saints' Day, which was solemnly observed; and as lights were burning upon the altars, great fires ensued, destroying what the earthquake had left. There were in all twenty-two shocks, and in Lisbon, or the immediate district, 60,000 lives were sacrificed. The action of the sea was not wanting to increase the horrors of the time. The waters at first retired from the harbour, leaving the bar

T

dry, and then returned in a wave 50 or 60 feet high. The main centre of this explosion was under the Atlantic, and the whole of Western and Southern Europe, including the British Isles, felt, in some degree, the force of the shock. In this country the lakes, ponds, and canals were most affected, and the waters of the Scottish lakes were disturbed. The sea wave raised by an earthquake has wide-reaching

Fig. 107.
View of the great Basalt-plain of the Snake River, Idaho, with recent cones.

effects on account of the mobility of water. All along the Spanish sea-board the waters swept onward with violence, at Cadiz destroying a sea-wall, and seaports on the opposite African coast were flooded. Far out at sea ships felt the shock, and along the British coasts the impact was felt even as far as Cork and Kinsale Harbours. ·

The American band.   The American band of volcanic activity stretches through the western sea-board almost from pole to pole

FIG. 108.—Mount Rainier—a Volcano, Puget Sound.

—from Tierra del Fuego or the "Land of Fire" in the extreme south, to the Arctic circle. A result of former igneous activity is represented (Fig. 107). The volcanoes are remarkable for their general energy rather than for their lava streams, whilst the earthquakes are of the most destructive kind. A mere enumeration of them would not be interesting; but the earthquake of Peru and Ecuador, in the year 1868, is supposed to have been one of the most destructive ever recorded. For 2000 miles along the Pacific sea-board the land was shaken; many cities and towns were ruined, and in a somewhat thinly-populated district thousands of people perished. The sea was, however, the chief destroyer, as at Lisbon. The waters at first retired, and, gathering force, returned in a mighty wave, carrying the wrecked vessels inland,[1] and inundating the already half-ruined towns and cities. One report describes the sea as rising like a mountain-side. The sea wave, after carrying destruction to the American coasts, passed right across the Pacific before coming to rest, travelling, it is computed, half round the world.

The Asiatic band.   The Asiatic volcanic band borders Eastern Asia, reaching from the Aleutian Isles to the Moluccas. The Aleutian Archipelago stretches like a bridge of fire from the Old World to the New, and is continued through the peninsula of Kamtchatka, which has over a dozen active volcanoes rising two or three miles above the sea. Through the Kurile Islands the volcanic band extends into Japan, and through the Philippine Islands to the East Indies and Java, which is the centre of perhaps the most tremendous volcanic pheno-mena in the world. Java alone has about forty vol-canoes. In one of the East Indian islands (all of which are volcanic), named Sumbawa, is the volcano of Tomboro, which in the year 1815 was in eruption for months. Great whirlwinds swept off houses,

[1] An American war-vessel was carried a quarter of a mile inland.

cattle, and people; lava and ashes covered the island, the dust being in such prodigious quantity as to darken the sun 300 miles away. Earthquakes were felt all around for 1000 miles, and the sea was covered with ashes. One part of the island permanently sank, the sea occupying its place, and out of a population of 12,000 but a miserable score survived. No words could picture the destruction which necessitated the alteration of the map of the district; and, subsequent to this catastrophe, other outbursts have occurred of almost equal violence, notably the one of which the volcano of Krakatoa was the centre.

Besides these regions of the first magnitude there are several minor ones where the earth's fires make themselves felt. Africa has no well-marked volcanic region, but the islands around its coast are generally volcanic. Iceland is an independent area with numerous volcanoes occasionally vomiting lava streams of enormous volume. New Zealand (Fig. 108) and the Sandwich Islands in the North Pacific are other regions of great energy. The island of Hawaii, the largest of the group, seems formed of igneous rocks, and has three remarkable volcanoes—Mouna Koa, Mouna Loa, and Kirauea. The first two named are of the usual type, and of great size and power, but Kirauea is a pit crater. It is oval in form with almost perpendicular sides, and is a striking permanent exhibition of igneous force. It is a vast cauldron of liquid fire, boiling and rolling in great waves, and with such a roar that, according to a visitor, "all the steam-engines in the world would be a whisper to it."[1]

The igneous forces, though bringing ruin and desolation upon man and his works, are yet to be regarded as part of the conservative agency of the world, by

Minor igneous bands.

[1] In the *Voyage of the Sunbeam*, where it is minutely described, Mrs. Brassey declares, "I could neither speak nor move at first, but could only stand and gaze at the horrible grandeur of the scene."

which the surface waste is renewed from the interior, and the balance held against the destroying powers of frost, rain, weathering, and marine denudation. The earth's fires are thus part of the orderly work of nature, and to them mankind is largely indebted for maintaining the world in its present habitable state.

XII

"YE SHOWERS AND DEW"

"Oh the rapture of beauty, of sweetness, of sound,
That succeeded that soft gracious rain !
With laughter and singing the valleys rang round
And the little hills shouted again.

"The wind sank away like a sleeping child's breath
The pavilion of clouds was upfurl'd ;
And the sun, like a spirit triumphant o'er death,
Smiled out on this beautiful world."

Southey.

THE water vapour in the air we breathe is of the first
importance in relation to health, whether invisibly absorbed or changed into dew, rain, or cloud; and, in its
condensed forms, adds greatly to the attractiveness of
nature. The beneficent influence of dew and rain was
early appreciated, as many Scriptural allusions make
clear. Allowance must be made for the fact that in
Bible lands the climate is hot and dry, and that, instead
of the wet weather common in the British Isles,
Palestine had only its "former" and its "latter"
rain, vegetation being refreshed by copious nightly
dews.

Hence dew is mentioned as a choice blessing in Dew a
the patriarch's prayer: "Therefore God give thee of blessing.
the dew of heaven, and the fulness of the earth,

and plenty of corn and wine"; and similarly in the benediction: "Blessed of the Lord be his land for the precious things of heaven, and for the dew." Silently and graciously bestowed, the sacred penmen find in it a fit emblem of the Divine counsel and help: "My doctrine shall drop as the rain; my speech shall distil as the dew, as the small rain upon the tender herb, and as the showers upon the grass"; or, "As the dew of Hermon, and as the dew that descended upon the mountains of Zion; there the Lord commanded the blessing, even life for evermore." The withdrawal of dew was regarded as a curse, as indicated in David's touching elegy upon his friend Jonathan: "Ye mountains of Gilboa, let there be no dew, neither let there be rain upon you, nor fields of offerings"; and its importance is shown in the solemn announcement of Elijah when before Ahab the king: "As the Lord God of Israel liveth, before whom I stand, there shall not be dew nor rain these years, but according to my word."

Scriptural
allusions.

In modern times, and by English writers, dew has been regarded as a pleasing accompaniment to the "miracle of morning":—

> "When many a pearly diadem
> Was hanging upon the rose's stem:
> And the fair lily's bell was set
> With a bright dewy coronet."

Iridescent
colours.

Dewdrops are nature's jewels; spangling in the morning light they rival the diamond's brilliancy. An observer, taking an early walk in the fields, and standing with his back to the rising sun, may perceive around the shadow of his head, as projected upon the grass, dewdrops giving forth shades of pale blue, brilliant white, straw-colour, pink, orange, and red, the drops five or ten yards distant from him exhibiting the play of two or three colours, those at a distance of fifteen or twenty yards showing the complete gamut of tints seen in the rainbow.[1]

The cause of dew, long a matter of speculation, has been made clear enough by the experiments and researches of several eminent English and foreign men of science. But, as clearing the way, it may be mentioned that dew does not descend like rain from the sky, and Wordsworth's couplet—

> " The dew was falling fast,
> The stars began to blink "—

must be taken as a poetical license. It does not rise like an exhalation from the ground, neither is it to be regarded as the perspiration of the vegetable kingdom, as once fancifully thought, although, of course, there is an ascent of dew-making material from the earth and from vegetation under the sun's heat. *The cause of dew.*

There are two kinds of dew, for which in France there are distinct names. There is a general production of dew or *serein*[2] throughout the air in the evening, and there is the special deposit of water in minute globules, as found upon grass and leaves. The second kind of dew can easily be produced by bringing into a warm apartment a decanter of ice-cold water, when there will be formed upon it[3] a film of water which is dew. That dew does not fall like rain is proved by the circumstance that the under surface of leaves is as much bedewed as the upper surface. The silent and invisible production of dew may be roughly illustrated by charging a large sponge with water to saturation. There is a *Experiments upon dew.*

[1] The order of the colours in the rainbow is violet, indigo, blue, green, yellow, orange, and red, the initials of which compose the word VIBGYOR. The reader may try how far, by gently turning his head, he can evolve these colours from a drop of dew that the sun shines upon.

[2] The Spanish name for dew is *serena*, from which we have the word *serenade*—music played out of doors when the dew is forming.

[3] Dew is also unpleasantly formed upon spectacles when the wearer enters a hot and crowded public building from a freezing temperature outside.

definite limit to its holding capacity. When quite full the least compression causes water to pour out of it. In like manner the atmosphere has a definite capacity for absorbing water vapour. Its carrying power for water depends upon its temperature ; the higher the temperature the greater the capacity for holding water vapour. Professor Ansted points out " that the quantity of vapour existing in air in the invisible form depends on local temperature, and is probably at no time uniform over any considerable area, and certainly never the same for twelve hours together in the same place. It is found that when the temperature is 50° Fahr. each cubic yard of dry air (about 168 gallons) can

Dew
evolved
from the
air.

hold a third part of a fluid ounce of water ; at 32° Fahr., or freezing point, only half this quantity is contained ; and at 70° Fahr. nearly double can be absorbed and retained in an invisible form." Thus, as the sun gains power, during the daytime, the absorbing capacity of the air rises higher and higher, until at noon, or shortly afterwards, its maximum temperature and carrying power for water vapour are reached. When, towards evening, the temperature is sensibly reduced, the air can no longer bear its burden of water vapour, but deposits it in the form of dew. Invisible when in the air, it shows as dew when condensed upon the objects around. The phenomenon is thus a clear case of evolution from the atmosphere, and wherever it penetrates near the ground dew is formed.[1]

The kinds
of dew.

Besides this general deposition of dew, caused by the reduction of the air temperature and its consequent inability to carry at night the water vapour acquired in the heat of the day, there is a second kind of dew,

[1] Longfellow's description of the formation of dew, in "The Song of Hiawatha," is scientifically correct :—

"Came as silent as the dew comes,
From the empty air appearing,
Into empty air returning.
Taking shape when earth it touches ;
But invisible to all men
In its coming and its going."

which is extracted from the air by bodies in contact
with it, which cool rapidly and to a large extent.
When, after a warm day, the sun's rays are withdrawn
from the earth, objects in nature radiate at very different
rates. The leaves and grass become cool straightway,
whilst rocks and water retain, long into the night,
some of the heat acquired in the day. Water especially
cools very slowly, and at midnight the temperature of
our rivers, lakes, and canals is nearly as high as at mid-
day. There is thus a heavy deposit of dew upon all   Pheno-
bodies radiating quickly, and it is wisely ordered that   mena of
foliage, which loses heat rapidly, is abundantly bedewed   dew.
and refreshed, whilst the gravel paths, the walls, and
the buildings, which would not be benefited by water-
ing, receive scarcely any dew. Experiments with the
thermometer illustrate the great reduction of temperature upon the grass as compared with that of the air
a few feet above, which will account for the copious
deposit of dew upon it.

The general phenomena of dew in this country may
be summarised thus :—

(1) Dew is most abundant at the equinoxes, spring
and autumn, for, at these seasons, the mid-day temperature is pretty high, causing the air to absorb moisture
extensively, and the nights are long and cool, causing
marked reduction of temperature, which is a condition
of copious deposition. In autumn the air is often
calmer than in the spring, which further conduces to
the formation of dew. In summer the air acquires a
higher carrying power for water vapour, but the nights
are then too short to permit of the necessary reduction of temperature. At mid-winter, on the other
hand, the nights are long and cold, but the air in the
daytime is not sufficiently heated to take up any large
store of dew-making material.

(2) In towns the streets and buildings radiate
slowly the heat acquired during the day, with consequent little loss of heat, and the dewfall is not per-

ceived; but in rural districts the grass and tree-covered surfaces rapidly lose their heat, and, as a consequence, are extensively bedewed.

(3) The human body receives but little dew of either kind, as it maintains a temperature of nearly 100° Fahr.

(4) A calm and windless night is favourable to its deposition, a breeze springing up reabsorbing any dew that may have been deposited; but a gentle air from the sea, as the north wind blowing upon Egypt, is favourable.

(5) Also a cloudless night, when radiation can proceed apace, is conducive to abundance of dew, whilst, when the sky is overcast, the escaping heat is beaten back and entangled between the clouds and the earth.

(6) The amount of dew annually registered in England varies from 3 to 5 inches, whilst the rainfall may be taken at 30 inches. The moon when shining clearly is thought to aid the dewfall; but this is erroneous, except in so far as a bright moon indicates free radiation of heat from the earth.

Injurious effects have been attributed to dew. The night must be cold for dew to be formed, and the lowered temperature and the change of water vapour from the absorption state to the solution state might be supposed to be harmful to the human constitution. This would be intensified in such countries as Egypt, Palestine, and Syria. In England, when the summer is exceptionally dry and hot, the refreshment of vegetation from the dews must be considerable. Hoar frost, whose effects are so beautiful, is simply frozen dew, when the ground temperature is below 32° Fahr.

The temperature at which dew begins to form is called the *dew point*. Below this reading the vapour of the air becomes condensed. The dew point is ascertained by means of the hygrometer, a simple form of which is the wet and dry bulb thermometer placed together upon a common stand. The wet bulb instru-

The hygrometer.

ment, which has a piece of muslin wrapped round it,
kept wet from a small vessel below, indicates, in
ordinary weather, a temperature lower than the dry
bulb, owing to the cold of evaporation depressing the
mercury [1] of the wet bulb. The difference between the
temperature shown by the dry thermometer and the
wet one indicates the humidity of the atmosphere.
In England the two readings are often close together,
and the depression rarely reaches 20°; but in some
tropical districts it is occasionally 50° or 60°, marking
an uncommon degree of dryness in the air. In several
hygrometers the temperature of the dew point is
indicated; but in the simpler contrivance of the wet
and dry bulb thermometer [2] the difference between
the two readings must be multiplied by a factor
supplied in the tables. The examination of the wet
and dry bulb readings may furnish a prognostica-
tion for the weather. To the invalid the amount of
dampness in the air may be important; and in the
greenhouse, and also in several manufactures, the
proper hygrometric state of the air may be essential
to the success of the operations. Although the
presence of a considerable amount of dampness in the
air is not injurious to health, it is desirable to bear in
mind that the great humidity of the English climate is
one of its drawbacks, and the "airing"—drying would
be a better term—of unworn clothing and unused
beds should be attended to with the greatest care.
Warning is given of unusual dampness in the air by

[1] To experience the cold of evaporation place a drop of spirits of
any kind upon the back of the hand. Waving the hand through the
air, a feeling of cold will be developed. The more rapid the evaporation the greater the depression of temperature.

Suppose the dry bulb to read 63° Fahr., and the wet bulb 54°
Fahr., the difference would be 9°, which, multiplied by 1·85 (the constant in this case to be used), would give 16·65. Subtracting this
amount from the temperature of dry bulb = 46·35° the dew point.

[2] The action of this instrument will be made more generally known
shortly, owing to a recent Act of Parliament requiring its use in weaving
and spinning factories to indicate the degree of moisture in the air.

Humid climate of England.
the long trails of steam from the railway engine ; by the dampness of rock-salt and seaweed hung upon the wall, and by the movement of the figures in the "weather house," a philosophical toy which is going out of use.

Dew and rain are associated phenomena. In this country, as has been explained, the dew deposited is only about a tenth of the annual amount of rain, and hence English farmers speak of the latter almost exclusively as affecting the crops. The deposition of dew is effected by the slow cooling of the air ; but when a stratum or stream of air laden with water vapour is suddenly cooled, rain may at once result, the amount depending upon the completeness of the change.

Formation of rain.
The great cause of rain, as of nearly all work in nature, is the sun. Day after day from the land, and the sea, the solar energy is drying up and evaporating water. In tropical lands, and in the hottest seasons the action is the most marked, and it is in such countries and seasons that the heaviest rains occur, with storms of thunder, lightning, and hail ; but in the ascent of rain material there is no sound heard—no roll of nature's machinery—and there is therefore danger of the operations being under-estimated. It is computed that the lifting power of the sun upon the waters is annually equal to 16 feet of water for the whole surface of the earth. The water rises invisibly until it reaches an elevation in the air where condensation sets in, and then the water vapour shows as a cap of cloud upon an invisible column of ascending vapour. Should the rain-bearing atmosphere suffer still further condensation, then the drops become too large for suspension in the air and there is rain.

The distribution of rain is according to laws which are now clearly demonstrated (Fig. 109). The heaviest rainfall will, of course, be in the tropics, where the annual temperature ranges from 80° Fahr. to 85° Fahr. In Guiana, Brazil, and North and Western India the average reaches 300 inches. Proceeding from the

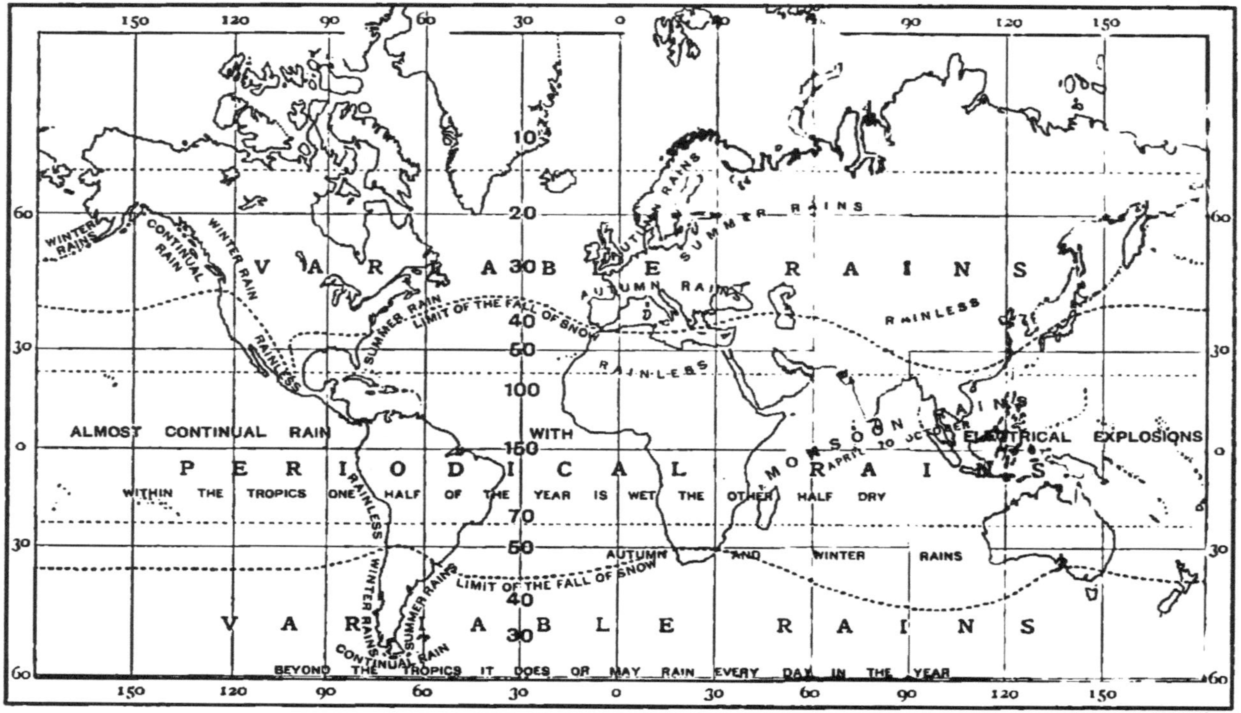

Walker & Boutall sc.

FIG. 100.—Distribution of Rain.

Distribu-
tion of
rain.

equator, north and south, the rainfall decreases with increase of latitude, the degree of which may be taken inversely to represent the evaporating power of the sun. A band extending over 4° of north latitude, and whose southern edge nearly touches the equator, is one of almost continual rain : that is, one in which rain occurs nearly every day, generally after noon ; and, in this region, the annual rainfall averages 150 inches or over 12 feet. At and near the Tropic of Cancer 100 inches are registered ; in latitude 30° north, 50 inches may be taken ; and, in the parallel of London, about 30 inches, which amount fairly represents the average of the British Isles.[1] In high northern latitudes the rains are diminished, and around the poles precipitation is in the form of snow. The southern hemisphere may be regarded as the great reservoir of rain ; but the fall of rain is greater in the northern hemisphere. At the Tropic of Capricorn it is 70 inches ; at 30° south 50 inches, and at the 50th parallel north and south 30 inches. But whilst in equatorial countries rains occur with a violence unknown in extra-tropical ones, it is neverthe-less a fact that the great rainless bands of the earth are in tropical or sub-tropical regions far from the sea. This is owing to one of the conditions for abundant rain—a good condensing medium—not being present.

Rainless
regions.

In the Old World is an almost continuous rainless band, reaching from the Atlantic Ocean to the Pacific, occupy-ing from the 4th to the 15th parallel of north latitude, and from the 16th meridian west to 118° east, and including the Great Desert of Sahara, the Arabian deserts, Persia, several of the countries of South Central Asia, and Mongolia. Rain - bearing winds sweep over these arid districts without having their vapour condensed, the hot sandy soil producing the

[1] District.	Rainfall in inches.
Malabar, coast of India . . .	135
England (average)	31
St. Petersburg	16

opposite effect of rarefaction, viz. a condition under which rain is impossible, and districts far from the sea would receive the winds in a dried state. In the New World the rainfall is throughout heavier than in the eastern hemisphere, and there are no very extensive rainless regions. In South America west of the Andes, and close to the Pacific, are several narrow coast regions where very little rain falls. In Mexico is a long narrow district of this kind, and there is also a limited area in the extreme north of South America, and almost contiguous to Brazil, which, however, is a country of tremendous rainfall. England has a rainy climate if estimated by the number of wet days, but this is by no means an infallible guide to the total rainfall, the number of wet days increasing generally with distance from the equator; but local conditions have a marked influence. The verdure of the English landscape is a consequence of the fact that England has about 150 days upon which rain falls; and similarly the grass-covered surface of Ireland has justly earned for it the name of the Emerald Isle. If rain is sometimes an inconvenience it must be remembered that its character and amount, even in England, constitutes one of our physical advantages—an estate, for example, being valuable in proportion to the rainfall which it receives.

Distribution of rain.

In the British Isles, Ireland, as first encountering the rain-bearing winds from the Atlantic, has heavy rains. In Scotland, which has an insular position, no part being far from the sea, the rains are also frequent and heavy, some of the effect being due to the height of the country, the air of elevated places forming a powerful condensing medium. On both coasts the influence of the Scottish hills is very marked. Glasgow, in the open, has 33 inches of rain, whilst Greenock, which is closer to the mountains, has 52 inches. In England the western coasts and the counties contiguous have abundant rains. The west of

Rains of England.

U

England is, consequently, largely devoted to pasturage
and dairy work, the cream of Devonshire and the cheese
of Gloucester and Cheshire having high repute. On
the eastern side of the country, and in the Midlands,
the rainfall is diminished, and the conditions are more
favourable for tillage and cultivated crops.[1] The
eastern counties largely produce corn, the south-eastern
counties hops and fruit. Elevated regions lying to the
west have abundant rain, as the Lake district, the village
of Seathwaite, which has sometimes nearly 200 inches
of rain, being the wettest village in England, whilst
Kendal and Keswick are two of the wettest towns. In
England sustained rain of tropical violence is unknown,
the greatest amount recorded in the twenty-four hours
being under three inches.

The
English
rainfall.

Raindrops vary in size from a twenty-fifth to a
quarter of an inch in diameter. The fall is accelerated
according to the law of gravitation, but ultimately
the descent becomes uniform owing to the resistance
of the air. The rate of movement is sometimes
half a mile per minute; but when frozen as hail the
descent is greater, sometimes reaching a mile a minute,
a velocity accounting for the destruction wrought by
hailstorms, as observed in that at Vienna in June
1894, when thousands of houses had their windows
broken, the hailstones in some cases perforating glass
like a bullet.

The rivers of a country are natural rain-gauges,
disastrous floods resulting from heavy rain, and which are
increased rather than mitigated by improved drainage.
The heavy rains falling upon the eastern slopes of the
Rocky Mountains occasionally swell the feeders of the
Mississippi, producing floods of destructive power. An
American naturalist has described the steady rise inch
by inch for weeks (the flood lasting sometimes for two

[1] The mean rainfall of the British Isles is 31 inches. At Seathwaite
the average is 127 inches; Dublin, 22; Edinburgh, 25; Liverpool,
34; Leeds, 21; London, 24; Manchester, 34.

months); the outflow of the stream on each side into
the low-lying country; the removal of the inhabitants
to the highest ground, rafts being ready should the
only means of conveyance be by water. Here and
there forests are undermined, trees plunging into the
foaming river, and cattle, horses, bears, and deer are
seen swimming for their lives. After the waters
subside the topography is altered. New streams
appear, and the main stream is seen to have altered
its course. The river is here the dominant power, and
all the works of man seem feeble beside it.

Natural rain-gauges.

Professor Ansted estimates that the total rainfall
upon the land only is not much less than 200 millions
of millions of tons annually. The solar action is of
course continuous, for behind the clouds, as in an
overcast country like England, the sun is constantly
putting forth its might.

Besides its reviving effect upon vegetation abundant
rain purifies the air, washing into the soil organic
and inorganic pollution. That England is a healthy
country is largely due to the number of its rainy
days. Wet seasons are proverbially healthy, whilst
in continued bright sunshine the air becomes filled
with impurities.

Blessings of the rain.

The clouds may be regarded as intermediate between
the invisible vapour which pervades the air, and rain.
In this country the clouds, although seldom calling
forth remark, except from closely observant persons or
those of artistic tastes, are nevertheless objects of rare
beauty. The poets have set forth their appearance in
attractive verse, and the English school of landscape
painters is largely indebted to cloudland for some of
the finest effects. Special forms and colours are
portrayed in the "Gainsboro' sky"; and Mr. Ruskin
has with pen as well as pencil endeavoured to show
forth their beauties, ranging from the light fleecy
clouds of the upper air to the heavy masses of cumuli
which may develop into storm clouds. "Leagued

Clouds in English landscape.

leviathans of the Sea of Heaven ; out of their nostrils
goeth smoke, and their eyes are like the eyelids of the
morning.   The sword of him that layeth at them
cannot hold : the spear, the dart, nor the habergeon.
Where rise the captains of their armies ?   Where are
set the measures of their march ? "

The clouds   As to the precise manner in which clouds form and
a difficult   float in the air, the inquiry of the ancient writer,
study.   " Knowest thou the balancings of the clouds ? " is still

Fig. 110.—Cirrus.

pertinent.   They possess the shifting, formless character of vapour, becoming visible or remaining invisible
according to minute changes of temperature in different
parts of the great aerial ocean ; they combine distinct
outlines with striking colours and other characteristics,
enabling us to speak of them as definite objects, with
Why   a capacity for change that is almost unlimited.   Their
clouds   buoyancy is somewhat difficult to explain ; they do not
float.   float upon the surface of the air like boats on water,
but are stratified in its mass.   They are not suspended
like feathers or other light objects, neither do they
float like balloons ; nor does cloud, like smoke or

incense, directly ascend under the influence of heated
air. The vapour atmosphere, though associated with
the air, is not in chemical union with it, and the gases
composing it, and probably the vapours in it, diffuse
themselves according to the established but curious
law by which the heaviest and the lightest intermix
and change their elevation.

Clouds are formed at various heights. They frequently occur at the elevation of four or five miles, and

Fig. 111.—Cumulus.

the lightest even higher. Observers two miles above the
sea have seen clouds far overhead, whilst, on the other
hand, rain clouds are often so little elevated as to trail
upon the tops of mountains not 1000 feet high. The
height of the clouds can be measured by the theodolite
of the land surveyor, and when frozen may be calculated roughly by taking advantage of the known rate
of cooling of the air until the snow-line is reached. The
shapes of clouds are almost endless, but there are
certain well-known types.

There is the cirrus or curl-cloud, resembling a curled   Kinds of
lock of hair (Fig. 110), which sometimes forms a beauti-   clouds.
ful network, or appears like flakes of snow. These

clouds, called "cats' tails" or "mares' tails," are the highest of all, and must generally consist of frozen vapour.   When the air is dry the cirrus-cloud has sharp edges, but in rainy seasons the definition is less clear.   Their height is always great—from three to five miles—Mr. Glaisher, in one of his scientific balloon ascents, when four miles high, having seen cirrus-clouds high above him.

Another characteristic cloud is the cumulus (Fig.

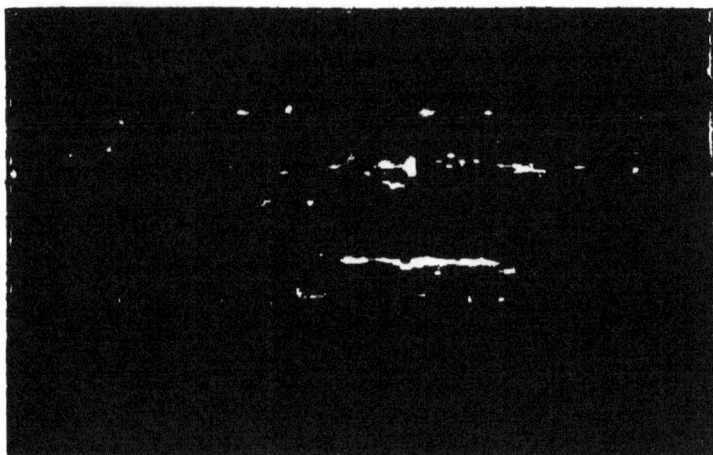

FIG. 112.—Stratus.

Cumulus and other leading forms.

111), or, as it is also called, the stacken-cloud.   This, the cloud of morning, is firm and solid-looking, and has a horizontal base where the ascending vapour begins to be seen, increasing like an irregular stack on the upper surface.   This form of cloud has often a noble appearance, and not unfrequently shows a silver lining.   The contrast of its pearly colour with the azure of the sky is occasionally one of the beauties of landscape.   It is often a mile high, and may be piled to the height of a quarter of a mile from the base.·   Such clouds, besides their general uses, serve as reflectors for the sunlight, and also shade the earth from the intensity of the sun.

The stratus, or fall-cloud (Fig. 112), is characteristic of the night. Its elevation is very slight, not unfrequently resting near the surface of the land or the water, and it may often be seen as a low-lying mist in valleys traversed by a stream. On the appearance of the morning sun the stratus-cloud rises irregularly and is absorbed into the atmosphere.

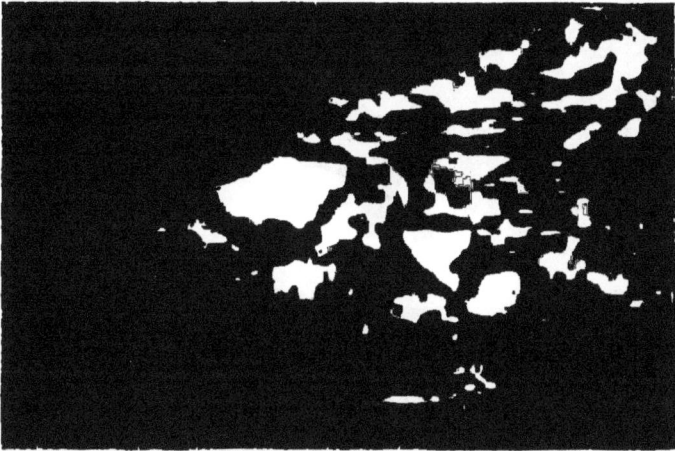

FIG. 113.—Nimbus.

Other minor forms of clouds are produced by a union of the primary ones, as the cirro-cumulus or " sonder-cloud." These are the clouds of calm summer weather, and their appearance, when separated in a regular manner, has been likened to a flock of sheep.

> " Scattered immensely wide from east to west,
> The beauteous semblance of a flock at rest ;
> These, to the raptur'd mind, alone proclaim
> The mighty Shepherd's everlasting name."

The cirro-stratus, or wave-cloud, has often a long

horizontal reach, and seems composed of fibres closely woven together.

The cumulo-stratus has often a threatening appearance, justifying the appellation of "war-cloud," and may develop into the rain-cloud or nimbus (Fig. 113), which is frequently an aggregation of all the forms of

FIG. 114.—Cloud Scenery.

cloud into one mass of pretty uniform appearance and oftentimes of enormous extent. Sometimes the whole of England is covered with these clouds, and their thickness is upon a commensurate scale.

It is upon the sombre nimbus-cloud that the rainbow is formed, and the iridescent hues of nature's triumphal arch are brought out in stronger relief by the dark background which serves as a foil.

Fig. 115.—Cloud Scenery—The Firth of Tay.

During stormy weather portions of vapour are torn
away from the general mass, and sail rapidly through
the air at a lower level. These are known as "the
flying scud." During great tempests the clouds assume
a threatening aspect, as in the storm which Dickens
has depicted in *David Copperfield* as bursting upon the
eastern coast : "Referring to the alarming state of
the sky the coachman replied, 'That's wind, sir, there'll
be mischief done at sea, I expect, before long.' It was
a murky confusion—here and there blotted with a
colour like the colour of the smoke from damp fuel—
of flying clouds tossed up into most remarkable heaps,
suggesting greater heights in the clouds than there
were depths below them to the bottom of the deepest
hollows in the earth, through which the wild moon
seemed to plunge headlong, as if, in a dread disturbance
of the laws of nature, she had lost her way and were
frightened. There had been a wind all day; and it
was rising then, with an extraordinary great sound.
. . . Sweeping gusts of rain came up before this storm,
like showers of steel ; and at those times, when there
was any shelter of trees or lee walls to be got, we were
fain to stop, in a sheer impossibility of continuing the
struggle."

Storm-
clouds.

Clouds play a useful part in the economy of nature.
They are the water-carriers of nature, conveying moisture from the sea, the great storehouse, far inland ; they
indicate the play of the electric forces, and they diffuse
the solar light and heat.

Cloud
photo-
graphs.

The study of clouds has been advanced by means of
photography (Figs. 115, 116, and 117). Instantaneous
pictures of storm-clouds have been preserved, and meteorological science has in this way been assisted. As weather
prognostics clouds have been carefully observed, and seafaring men and fishermen trust to their indications as
well as to those of weather-registering instruments.

"In England mares' tails portend wind," according
to Abercromby, "and goat's hair only rain"; while

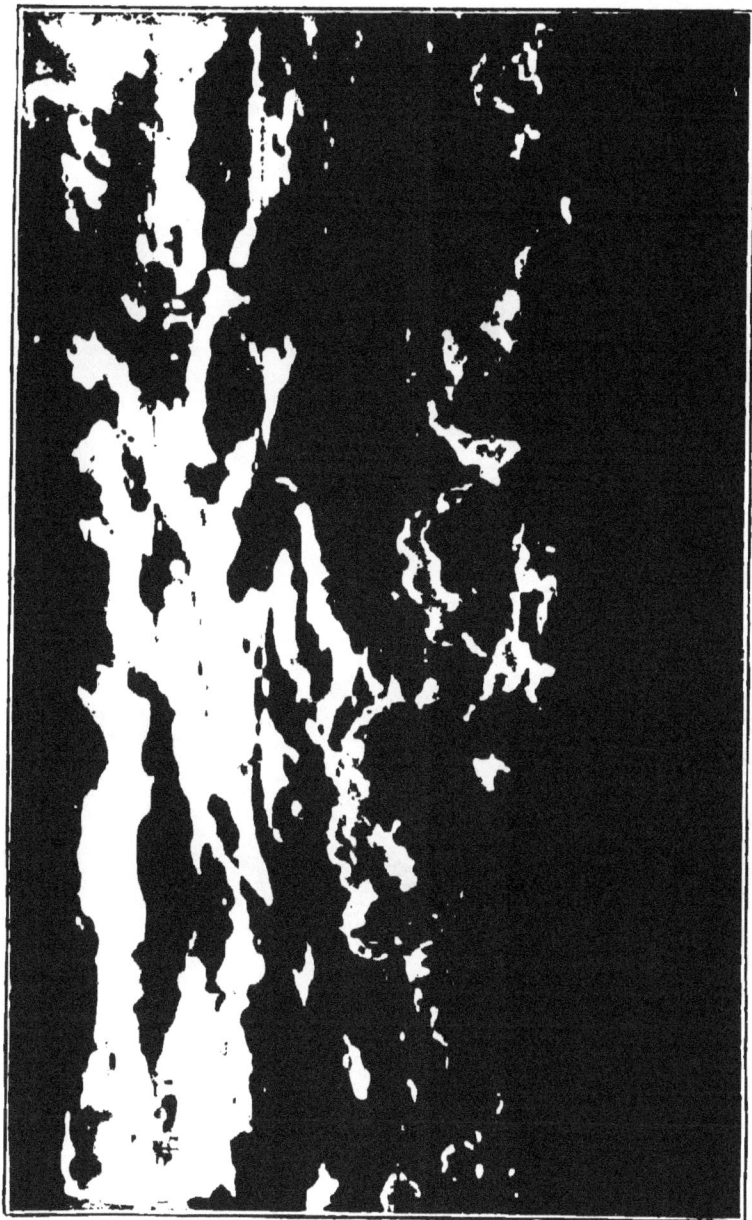

FIG. 116.—Cumulus and Strato-Cumulus.

Cloud prog- nostics.
"mares' tails and cats' tails precede every hurricane in the tropics." The mackerel sky is a certain prognostic of fine weather. "Mackerel sky twelve hours dry" is an observation much trusted to in rainy Ireland.

Fig. 117.—On the Lima and Oroya Railway. Among the Clouds.

Clouds should be studied from nature as well as from books, as, according to one of the best authorities, "a few hours spent in watching the changing and degrading forms in a sky which is covered by detached cumulus, or the very different modifications almost from minute to minute of cirro-stratus, will better assist any one to understand the nature of cloud forms than reading pages of the best printed matter."

The study of nature in any department will give a freshness to life, and will furnish resources that add to human enjoyment; and in the wide field of scientific study nowhere than in the new study of meteorology is there more inviting material. "Between the earth and man arose the leaf. Between the heavens and man came the cloud: his life being partly as the falling leaf and partly as the flying vapour."

GLOSSARY

Alabaster (p. 35).—A marble-like mineral composed of sulphate of lime with some water ; of various colours—white, yellow, red, or gray ; named from Alabastron, a village in Egypt.

Amazon (p. 141).—The giant river of South America, named after the Amazons, a mythical race of warrior women originating in the Caucasian Mountains. The Spanish conquerors of South America spoke of armed women on the banks of this river and its tributaries.

Anacondas (p. 74).—The popular name of two of the largest species of constrictor serpents (whose forms are simulated in tropical forests, *vide* Fig. 28) ; sometimes reaching the length of 30 or 40 feet, often brilliantly coloured ; destitute of poison fangs, but killing their prey by crushing it.

Archipelago (p. 2).—Originally descriptive of the sea, studded with islands, which separates Europe from Asia ; now applied to any group of islands.

Aristotle (p. 3).—Born at Stagira in Macedonia B.C. 384 ; founder of the peripatetic school ; physics and the study of natural phenomena were amongst the subjects he taught.

Asteriae (p. 98).—A group of star-fishes, the rays being suggestive of stars.

Calamus (p. 175).—A species of small palms, the stems of which are the rattan canes of commerce.

Cañon or *Canyon* (p. 45).—From the Spanish for a tube, or funnel ; applied to long and narrow mountain gorges with perpendicular sides, somewhat resembling tunnels through the rocks.

Carvel or *Caravel* (pp. 8 and 29).—The name given to the small ships or large boats employed by Spanish and Portuguese

navigators in the fifteenth and sixteenth centuries; narrow
at the poop, wide at the bow, with a double tower at the
stern, and a single one at the bows; provided with four
masts, and a bowsprit, and with principally lateen sails.

Challenger (p. 90).—A screw corvette of 2400 tons displacement,
which was sent out under the Government Hydrographic
Department in 1877 as an exploring ship.   The guns were
removed to accommodate laboratories, workshops, and cabins.
The instructions were to examine into the depth, deposits,
temperature, and physical and chemical properties of the sea.
The smaller marine creatures were to be collected, and reported
upon, and systematic magnetic and meteorological observations were to be made.   The vessel had a crew of 240 men,
and the late Sir Wyville Thomson was the scientific director.

Cimmerians (p. 4). — A mythical people dwelling in the far
western boundary of the ocean; their land, according to
Homer, was enveloped in constant mist and darkness.

Crevasses (pp. 39 and 244).—Cracks in glaciers descending irregularly to great depths.

Cyclops (p. 256).—A mythical race of giants, the sons of Neptune
and Amphitrite, inhabiting Sicily, and regarded as assisting
Vulcan, the armourer-god, to forge thunderbolts for Jove;
described as having a single circular eye placed in the middle
of the forehead.

Cymry (p. 156).—The name given to themselves by the Welsh.
The etymology of this word indicates that their dwellings
were at the confluence of rivers, where would be good pasture-
age and facilities for communication by water.

Denudation (p. 82).—The act of removing the surface of a rock by
water or weathering.

Diatoms (p. 90).—Simple microscopic plants; formerly classed
as animalculæ; invested with cases of silex consisting of two
symmetrical valves often beautifully figured.

Doldrums (p. 222).—The sailors' name for the narrow zone of
ocean, near the equator, characterised by frequent calms,
squalls, or baffling winds; also spoken of as the "horse
latitudes" by the early navigators from the number of horses
that died of the heat when becalmed in the long journey by
sailing vessels.

Echini (p. 98).—Sea-urchins, the body covered with a case provided with movable spines or prickles.

Elysium (p. 4).—In the mythology of South Europe, the place assigned to happy souls after death.

Epiphytes (p. 74).—Plants fixed upon tropical trees, but which do not derive nourishment from the supporting host, as many of the orchids.

Fauna (pp. 150 and 252).—All the wild animals peculiar to a natural-history region ; the name adopted from Fauna, the Roman goddess of fields and cattle.

Flora (pp. 150 and 252).—The botany, or complete collection of wild plants natural to a district ; the name from Flora, the Roman goddess of flowers.

Hermetically sealed (p. 188).—Absolutely shut off from external percolation and internal leakage ; from Hermes, the Greek god of sciences and commerce.

Hieroglyphic (p. 136).—Written in characters, words, or signs not easy to decipher, and conveying information not intelligible to ordinary persons.

Hyperboreans (p. 4).—In mythology, the people who lived in the far north ; but who, protected from the severity of the north wind, enjoyed perpetual sunshine with abundance of fruits.

Isobars (p. 222).—Imaginary lines passing through places having the same height of barometer. Each day these lines are shown upon the *Times* weather chart, with the values indicated. They are reliable guides to the direction and force of the wind.

Lateen sails (p. 106). — Triangular sails extended upon a yard sloping at about half a right angle with the deck of the vessel ; used upon small vessels on the Mediterranean.

Metamorphic slate (p. 47).—Generally clay slate, changed and hardened by heat and pressure.

Moraine (pp. 249 and 250).—From the Italian *mora*, a heap of stones ; mounds of stones, sand, and other materials found upon the sides and surface of glaciers, often piled up to the height of many feet ; or deposited in the warm lower valleys where the glacier dissolves.

Narbonnaise (p. 106).—Narbonensis, or Galli Narbonensis, a province of the Roman Empire occupying the southern and south-eastern part of Gaul.

X

Nilometer (p. 116).—A slender graduated pillar placed in a well which communicates with the river Nile; divided into 24 cubits, each of 21·4 inches. 21 cubits is a beneficial rise, whilst one of 24 cubits causes undue flooding and disaster.

Olympus (p. 1).—The highest mountain in ancient Greece, and the fabled home of the celestials, with Zeus or Jupiter at their head ; and the fancied central point of the earth's surface.

Palæocrystic Sea (p. 253).—Expanses of Polar ice which remains unmelted year after year, some of it being estimated to be 100 years old.

Palimpsest (p. 82).—A manuscript where the original writing has been rubbed out, and the parchment again written upon.

Pillars of Hercules (p. 2).—Two mountains guarding the entrance to the Mediterranean, known as Calpyê and Abyla, one on the European side, and the other in Africa.   Calpyê is now known as the Rock of Gibraltar, and Abyla as Mount Hacho, or Apes' Hill.

Polypidom (p. 104).—The reef or structure formed by the coral polype ; oftentimes of enormous dimensions, the sea-front being sometimes 1000 feet deep, with proportionate length and width.

Pythagoreans (p. 3).—The sect of philosophers founded by Pythagoras, who taught the doctrine of the transmigration of souls, and who had an elementary notion of the true system of the universe.

Pytheas (p. 5).—A navigator of Marseilles who, in the second half of the 4th century B.C., visited the coasts of Spain, Gaul, and Great Britain, and furnished the earliest precise information of the western countries of Europe.

Ribbon roads (p. 76).—A name given in Nottinghamshire to the drives, chiefly in Sherwood Forest, where the wheel-tracks are cut through the turf, presenting, for miles, the appearance of a striped ribbon.

The Roaring Forties (p. 11).—The part of the South Atlantic Ocean between the 40th and 50th parallels of latitude, where gales and heavy storms are frequent, by which the progress of vessels rounding the Cape of Good Hope on the way to India was retarded.

Sargasso Sea (p. 9).—An extensive area of the North Atlantic occupied by the Gulf weed and other forms of marine vegetation. One of the most common derived its scientific name (*Sargassum bacciferum*) from the clusters of grape-like air-vessels by which it is buoyed up.

Scalds (p. 2).—Ancient Scandinavian poets, who composed songs in honour of eminent men and their exploits, and who sang their compositions on public occasions.

Serpentine (p. 153). — A rock composed of hydrated silicate of magnesia ; rather dark-coloured, with shades and spots resembling a serpent's body.

Thales (p. 4).—One of the founders of the study of philosophy and mathematics ; taught that water is the origin of all things ; lived about 500 B.C. ; made the first recorded electrical experiment with amber, the Greek word for which is *elektron*.

Thebes (p. 122).—The capital of Upper Egypt, and one of the most famous cities of antiquity ; stood upon both banks of the Nile ; was the central station for the worship of Ammon ; and, according to the Homeric legend, had a hundred gates.

Titanic (p. 104).—Enormous in size and strength ; pertaining to the Titans, who, in the Grecian mythology, were the twelve children (six sons and six daughters) of Uranus (Heaven) and Ge (Earth). They fought against and dethroned their father ; but, after a long struggle, were defeated and thrown into Tartarus.

Vikings (p. 85).—Sea pirates, chiefly coasting the Scandinavian peninsula, and often making descents upon Britain ; the name is derived from *vik*, a bay.

INDEX

AFRICAN lakes, 166
    exploration, 168
Agents of waste, 62
Air, the, carbonic dioxide of, 206
    Cavendish's researches, 202
    chemical composition of, 201
    early ideas concerning, 193
    elasticity, 200
    extent overhead, 198
    Flammarion's estimate, 193
    Galileo's puzzle, 195
    humidity, 199
    medium of light and sound,
        200
    mobility of, 200
    nitrogen of, 206
    oxygen of, 205
    Pascal's experiment, 194
    pressure, distribution of, 222
    refractive power, 201
    relation of pressure to eleva-
        tion, 199
    stratification of gases, 198
    Torricelli's experiment, 195
    water vapour of, 205
Albert Lake, 165
Alexander the Great, expeditions,
    6
Amazon, scenery of, 141
Amazons, the, 4
America, discovery of, 10
    waterfalls of, 134
Andes, 49
Antarctic ice wall, 248
Ararat, Mount, 41
Arctic travel, 254
Artesian wells, 179

Asiatic waterfalls, 134
Auvergne Mountains, 54
Avalanche, stone, 66

BANN, salmon leap upon, 132
Basalt plains, near Snake River,
    274
Ben Nevis Observatory, 51
Bifurcation, 140
Blizzard, the, 209
Blue Mountains, 47
British seas, 88
Brooke, Sir James, and East
    Indies, 14

CAÑON, the Bloody, 45
    the Grand, 45
Cape of Good Hope, discovery of,
    11
Cape of Good Hope doubled, 12
Cashmere, Vale of, 34
Caspian Sea, 160
Caucasian Mountains, 41
Cavendish, weighs the earth, 22
Caverns, 79
    of Derbyshire, 80
Cimmerians, the, 4
Climate, the English, 221
Clouds, kinds of, 293
    as prognostics, 300
    cirrus, 292
    cumulus, 293
    nimbus, 295
    scenery of, 296
    storm clouds, 298
    strato-cumulus, 299
    stratus, 294

Clouds, why floating, 292
Clyde, Falls of, 129
Colchis, land of, 4
Cold of great elevations, 39
Columbus, character of, 6
  voyages of, 8
Compass variation discovered, 11
Congo, the, 125
Coniston Water, 160
Continents, articulation of, 18
Cotopaxi, 49

DEAD Sea, 162
Derbyshire springs, 182
Dew, Scriptural references, 279
  colour effects of, 280
  cause of, 281
  kinds, 282
Dropping Well, Knaresborough, 187

EARTH, the, early notions, 1
  chiefly solid, 257
  crust of, 21
  internal state, 27
  magnitude of, 25
  revolution round the sun, 26
  rotation of, 25
Earthquakes, Bible references, 260
  superficial, 288
Elevation, how determining climate, 19
Elysium, 4
England, geological structure of, 62
  climate, factors of, 286
Ethiopians, the, 4
Euphrates, 114
European plain, 70
  waterfalls, 132
Eyre, crosses Australia, 14

FALLS, of Clyde, 130 ; Glomach, 130 ; the Tees, 130; Powerscourt, 130
Fires of the earth, early superstitions, 256
Forest, African, 72

Forest, early English, 74
  Sherwood, 76
Frost, agent of waste, 65
  decorative effects, 240
  at Niagara, 253
  expansive force, 231
  flowers and crystals of, 240

GAMA, de, as explorer, 11
Geneva, lake, 163
Geysers, 182
  of Colorado valley, 183
Glaciers, Alpine, 237
  Mer de Glace, 251
  moraines, medial, 249
    lateral, 249
  movements of, 250
  work of, 252
Gravitation, lateral, 24
Gulf Stream, the, 101
  of the Pacific, 101

HIMALAYAS, the, 3
Humboldt, general order of the earth, 1
  Grecian scenery described, 4
  travels, 228
Hurricanes, 14
Hygrometer, use of, 284
Hyperboreans, the, 4

ICE age in Britain, 235
  exercise, 236
  molecular constitution, 240
  moulding of, 234
  palæocrystic, 253
  physics of, 246
Icebergs, carriers of nature, 66
  force of, 244
  how formed, 234
  number seen by Parry, 242
  tabular form of, 243
Ice-floes, 244
Igneous bands, the lesser, 277
Isabella, Queen, assisting Columbus, 9
Islands, 104
  coral, 104
  uses of, 105

JACOB's Well, 175

KATRINE, lake, 148
Killarney, lakes, 147, 160
Kirauea, pit crater of, 277

LAKES, changes in, 157
classified, 154
colour of waters, 169
English, 158
formation of, 154, 155
freshwater, 162
harvest of, 153
of great transparency, 170
remarkable, 168
salt-yielding, 150
scenery of, 148
Scotch, 160
uses of, 148, 150
Land, elevation of, 20
horizontal distribution, 17
Land and sea breezes, 223
Landscape, English, 60
types of, 62
Lava streams, 267, 270, 271
Leichardt, Australian explora-
tions, 15
Livingstone, missionary journeys,
16
Lisbon, earthquake of, 273
Lochlands, how formed, 152
Lodore, Falls of, 128

MAGELLAN, voyage of, 12
Markham, Captain, Arctic jour-
ney, 18
Matterhorn, the, 40
Mediterranean, the, 106, 262
Midianites, the, 5
Mirror Lake, the, 170
Mississippi, river, 142
Missouri, river, 142
Monsoons, 225
Mont Blanc, 36
Mountains, Ben Venue, 30
beacons, used as, 56
breadth of ranges, 33
early ideas of, 31
general view of, 32
places of refuge, 31

Mountains, uses of, 50, 52
Murchison Falls, 135

NANSEN, Arctic voyage, 18
Niagara Falls, 134
Nile, the, 115
flooding of, 116
Norsemen, their ideas of the
universe, 2

OASES, 76, 78
Oceanus, river, 4, 100
Ohio, the, 120

PERSONIFICATION of natural ob-
jects, 58
Pitch Lake, Trinidad, 167
Plains, desert, 76
Polar seas, 107
exploration, 254
ice, extent of, 245, 247
Pytheas, early voyages, 5

RAFFLES, Sir Stamford, and East
Indies, 14
Rain action, 64
distribution, 287
formation, 286
Rainless regions, 288
Rains of England, 289
Reclus, Homeric view of earth,
1, 2
Regelation of ice, 234
Rhine, the, 119
Rhone, the, 119
Rivers, agents of waste, 68
basins of, 126
boundaries formed, 123
cities upon, 122
comparison with time, 113
flooding of, 116, 138
gauging the rainfall, 291
human life, 124
mouths of, 140
rivers, rapids of, 127
scenery of, 127
Scriptural allusions, 113
sources of, 124
Robin Hood's Well, 189
Rocky Mountains, 43

Roman Empire, growth of, 6
Rydal Water, 161

SALT Lake, the Great, 172
Scenery, early notions, 58
  mountains, 69
  reflected, 170, 172
  types of, 69
  variation of, 59
  words expressive of, 59
Schools, ancient, the Pythagoreans, 3
Sea, colour of, 97
  composition of, 93
  depths of, 88
  destroying action, 64
  early ideas, 84
  endurance of, 83
  English love of, 85
  extent of, 86
  floor of, 90
  harvest of, 109
  importance in nature, 83
  man's control of, 85
  phosphorescence, 98
  pressure at great depths, 90
  saltness of, 94
  shells, 108
  specific gravity of, 94
  temperature of, 99
  transparency, 98
  uses of, 109
Sierras of Mexico, 43, 48
Siloam, Pool of, 177
Sinai, 43
Snow crystals, 238
  line, 26
Springs, 175
  boiling, 182
  calcareous, 188
  ebbing and flowing, 179
  hot, 180
  medicinal, 188
  mineral, 183
  oil-yielding, 184
  perennial, 176
  reciprocating, 177
  salt, 186
  siliceous, 190
  sulphur, 186

Springs, uses of, 191
Stanley, exploring journeys, 16
St. Anne's Well, Buxton, 181
St. Lawrence, river, 142
Steppes of Russia, 70
Storm winds, 226
  of England, 226
Stromboli, 261
Sturt, discoveries of, 14
Sun, physical control of earth, 258
Susquehanna, river, 121
Swiss lakes, 152

TABOR, Mount, 42
Tanganyika, lake, 166
Thales, teaching of, 5
Thames, scenery of, 144
Trade winds, discovered by Columbus, 9
Trade winds, how caused, 214
  use in navigation, 48

UNDERGROUND waters of England, 177

VELINO Falls, 132
Vesuvius, Mount, 262
  eruptions of, 263, 269
Victoria Falls, 137
Volcanic bands, 258
  of America, 274
  of Asia, 276
  of Europe, 260
  eruptions, 263-269
  products of, 272
  submarine, 259
Volcanoes of New Zealand, 275

WATERFALLS, 128
Water-partings or water-sheds, 36, 127
Wave motion, 102
Weather, English, 220
  Ballot's law, 222
  work, 66, 207
Wells, custom of flowering, 190
  early mention of, 174
  Eastern value of, 175
  holy, 190

West Indian storms, 228
Wharfe, river, 127
Winds, anti-trades, 218
  causes of, 213
  chart of, 222
  local, 209
  direction of, 208
  early ideas of, 212
  English, 220

Winds, illustration of, 214
  seasonal, 225
  velocity of, 208
  ventilating the earth, 221
Windermere, lake, 158
Winifred's Well, 176

ZERMATT, glacier, 233

THE END

Printed by R. & R. CLARK, *Edinburgh.*

BOOKS FOR STUDENTS OF ASTRONOMY.

By J. NORMAN LOCKYER, F.R.S.

A PRIMER OF ASTRONOMY. Illustrated. Pot 8vo. 1s.

ELEMENTARY LESSONS IN ASTRONOMY. With Spectra of the Sun, Stars, and Nebulæ, and Illus. Thirty-sixth Thousand. Revised throughout. Fcap. 8vo. 5s. 6d.

QUESTIONS ON THE ABOVE. By J. FORBES ROBERTSON. Pot 8vo. 1s. 6d.

THE CHEMISTRY OF THE SUN. Illustrated. 8vo. 14s.

THE METEORITIC HYPOTHESIS OF THE ORIGIN OF COSMICAL SYSTEMS. Illustrated. 8vo. 17s. net.

STAR-GAZING PAST AND PRESENT. Expanded from Notes with the assistance of G. M. SEABROKE, F.R.A.S. Royal 8vo. 21s.

By HUGH GODFRAY, M.A.
Mathematical Lecturer at Pembroke College, Cambridge.

A TREATISE ON ASTRONOMY. Fourth Edition. 8vo. 12s. 6d.

AN ELEMENTARY TREATISE ON THE LUNAR THEORY. Crown 8vo. 5s. 6d.

————————

POPULAR ASTRONOMY. A Series of Lectures delivered at Ipswich. By Sir G. B. AIRY, K.C.B., LL.D., D.C.L., late Astronomer-Royal. Revised by H. H. TURNER, M.A., B.Sc., Chief Assistant, Royal Observatory, Greenwich. Seventh Edition. Fcap. 8vo. 4s. 6d.

POPULAR ASTRONOMY. By S. NEWCOMB, LL.D., Professor U.S. Naval Observatory. Illustrated. Second Edition, Revised. 8vo. 18s.

PIONEERS OF SCIENCE. By OLIVER J. LODGE. Extra Crown 8vo. 7s. 6d.

THE PLANET EARTH. An Astronomical Introduction to Geography. By R. A. GREGORY, F.R.A.S., University Extension Lecturer. Illustrated. Globe 8vo. 2s.

THE ROMANCE OF ASTRONOMY. By R. KALLEY MILLER, M.A., Fellow and Assistant Tutor of St. Peter's College, Cambridge. Second Edition. Crown 8vo. 4s. 6d.

MACMILLAN AND CO., LONDON.